PUBLIC SPEAKING AND TECHNICAL WRITING SKILLS

FOR ENGINEERING STUDENTS

PUBLIC SPEAKING AND TECHNICAL WRITING SKILLS
FOR ENGINEERING STUDENTS

Second Edition

P. Aarne Vesilind

Professor of Engineering, Emeritus
Bucknell University
Lewisburg, PA

Lakeshore Press
Woodsville, NH

iv

CONTENTS

A NOTE TO THE INSTRUCTOR

During the past decades it has become quite clear that communication skills continue to be a most important aspect of professional engineering, and the need to improve these skills across our profession has been recognized by the Accreditation Board for Engineering and Technology (ABET). Engineering communication skills differ from those in other professions and thus it makes sense to introduce these within the engineering curriculum.

This Second Edition of *Public Speaking and Technical Writing Skills for Engineering Students* has been significantly updated and revised in response to the needs of instructors who have used it in their classes. The most important changes are the inclusion of a major section on writing style and the use of PowerPoint software. Also new is a chapter on the skills necessary for conversing one-on-one with clients, reporters, lawyers, and prospective employers.

The last chapter of this book is a discussion on the ethics of engineering communication. In using this book for instruction, the ethics of communication can be introduced separately, or, as is the case with most instructors, the ideas in the last chapter can be infused throughout the discussions on speaking and writing. When I taught this material I used three class periods for the writing and speaking topics and introduced ethical concerns throughout the course. An alternate method is to introduce ethics immediately following the chapters on speaking and writing. Either way, ABET emphasizes that ethics represents a major part of an engineering education and has made this a requirement for accreditation. Introducing professional ethics with technical communication skills affords a useful and efficient way of covering both topics.

An instructor's manual for this book is available. Please contact Lakeshore Press, P. O. Box 92, Woodsville, NH 03785, telephone 603-747-8083.

ACKNOWLEDGMENTS

Recognizing that the ability to communicate often dictates success or failure of a professional engineering career, the late Professor Eric Pas of the Department of Civil and Environmental Engineering at Duke University and I developed a course for the engineering curriculum that covers technical writing, public speaking, engineering drawing, and computer graphics. For over ten years we spent many hours discussing the sequence and value of the material in this course. Some of his insights have no doubt permeated my own thinking and have therefore found their way into this book. I will always be grateful to Professor Pas for sharing his communication skills with me and for being the best friend a person could ever hope to have. It is thus with appreciation that I dedicate this book to his memory.

I finished the first edition of this book when I was on sabbatical leave at Bucknell University. The hospitality of Jai Kim, Chair of the Department of Civil Engineering, is very much appreciated.

Libby Vesilind, Priit Vesilind, Pamela Vesilind, and Christopher Endy read various versions of the manuscript, and as a result of their suggestions and criticisms the book has been greatly improved, and I have become, I hope, a better writer.

CHAPTER 1

MODES AND FORMATS IN ENGINEERING WRITING

Life is an essay question, not a True or False test! We are continually called on to explain in words our ideas, our wishes, and our emotions. Verbal communication is as essential in engineering as are calculus and mechanics, and learning to write and speak well is a prerequisite for a successful engineering career.

There are two reasons why engineers should learn to write well. The first, and most obvious one, is that writing is one of the engineering languages with which one person communicates information to a colleague or a client. For many years this was thought to be the sole purpose of technical writing – to write in such an unambiguous manner that the meaning could not be misunderstood. This is still a valid purpose, but it now joins a second more subtle but equally important reason for verbalizing information. The process of writing, or putting thoughts into words, is also a **learning** process for the writer. Writing crystallizes unorganized thoughts that have been rattling around in the head and this process of organization produces ideas and concepts that might not otherwise have been articulated. Thus good technical writing not only conveys information in an unambiguous manner but also serves as a means of thinking. Engineers often almost apologetically say, "I can't think without a pencil," and they use doodles as a graphical thinking tool. The same is true for verbal language and, the more developed the verbal facility, the better the thinking process.

In this chapter I discuss four different types of writing modes and then examine the formats or types of engineering documents usually encountered in practice. Following a short discussion of technical writing style, we look at the actual process of writing – the "how to do it" step.

1.1 WRITING MODES

Types of writing can be divided into an almost infinite number of categories. One useful definition of various types of writing is suggested by Houp and Pearsall.[1] According to them, all writing can be categorized as:

- exposition
- narration
- description
- persuasion

Most written documents contain some aspects of each category, and understanding each of these four can give you greater control over your writing.

1.1.1 Exposition

Expository writing clarifies, explains, instructs, puts forth, or exposes some material. Its purpose is to share information with the reader. Expository writing works best when the writer uses certain rhetorical devices already familiar to the reader, including topical arrangement, exemplification, definition, and comparison.

Topical Arrangement. To communicate a large amount of material, the writer can break down the substance of the material into an organized topical arrangement. For example, this book is broken down in such a manner, with individual chapters and subheadings. Imagine what this textbook would have looked like if it were a complete mishmash of all the information in a totally unorganized manner.

The procedure for achieving a proper topical arrangement is the *outline*. You probably have been writing outlines since the third grade, often without understanding the real purpose of these exercises. The outline is nothing more than a topical arrangement that makes the material easier to write and makes the written work more readily understood by the reader. To work best, the outline should have several necessary attributes:

- There should be a logical progression of ideas.
- Headings should be of equal weight.
- A series or headings should be of one classification.
- Topics should not overlap.
- Divisions should be of sufficient size to incorporate meaningful information.

Often a good method of topical arrangement is chronology; the use of time as a variable. In other cases space is useful. For example, if it is necessary to describe a flower garden, either time or space can be the variable. If time is used, the exposition starts with the first flowers to appear in the spring and moves through the seasons. If space is used, the garden at any given time can be described according to what flowers appear from one end of the garden to another. These are not necessarily the only topical arrangements, of course. A lovely description of a garden is possible by invoking neither time nor space as a vehicle but choosing color instead. If the purpose is to describe a garden, the writer should first imagine as many topical arrangements as possible and then, after choosing one, remain consistent.

Exemplification. Provide an explanation for statements

that leave some question unanswered. For example, the statement, "The grades in this course last year were too low," demands more information. How low? Compared to what other course? Please explain yourself. Exemplification is a technique where a general statement is made first, followed by explanations that provide the specific material. Without the explanations the statement is of little value.

Definition. One of the most difficult tasks in technical writing is the adaptation of the written material to the audience. If the audience is made up of one's colleagues, the language would be very different from the language used in an engineering report to the elected officials of a community. Which words or symbols need definition, and which do not? If the report includes the word "odometer," for example, should it also include the definition "an instrument in a car used for measuring distance," or is it possible to depend on the audience's knowledge of automobiles?

The reader's knowledge of symbols is also difficult to gauge. For example, the letter Q is universally (well, at least in the U.S.!) used as a symbol for flow rate. But only engineers would know this. So if flow rate is used in a discussion for non-engineers, the words "flow rate" should be spelled out, or if Q is used, it should be defined from time to time. Further, flow rate can be in many units (e.g. million gallons per day, cubic feet per second, and even some abominations such as acre-feet per year). The reader who is not educated in the discipline could not guess what MGD or cfs mean – terms familiar to most engineers. In writing definitions it is wise to err on the side of redundancy.

Comparison. Comparisons, especially analogies or metaphors, are powerful rhetorical devices. Visualizing electricity in terms of fluid flow is an old comparative device for achieving a better understanding of electrical current. Describing a fabric filter in an electric power plant as a big vacuum cleaner can be helpful in explaining the concept to a non-technical audience. Analogies compare one situation or thing to another, often using words such as "like" or "as." For example, "Cleaning house when your children are still home is like shoveling the sidewalk when it's still snowing," (Attributed to Phyllis Diller), or Calvin's observation that "One inch of snow is like winning 10 cents in the lottery," (*Calvin and Hobbes* comic strip).

Metaphors, on the other hand, invoke a situation that can be useful in attaining a deeper meaning but are not direct one-to-one analogies. For example, the sentence, "The chemical engineering curriculum is a salad bar of courses," uses salad bar to describe the curriculum. In ordinary language, salad bar describes one thing, but as a metaphor it is applied to another (the curriculum) without expressly indicating the relationship between the two.

In summary, expository writing conveys information using rhetorical devices such as a proper topical arrangement, adequate exemplification, useful definition, and effective comparison.

1.1.2 Narration

Why do many public speakers start their presentation with stories, personal experiences, or jokes? The reason is, of course, that stories or jokes tend to "warm up" the audience, to establish a rapport between the speaker and the audience.

Narration is the oldest form of information transfer. When we were young, we all listened to stories, and storytelling is a very enjoyable form of communication, even for adults.

Narration takes place when at least some of the following parts are present:

- time – "Once upon a time..."
- place – "in a land far, far away..."
- characters – "there lived a beautiful princess..."
- plot – "who wanted to study engineering."

Narration is seldom used in engineering writing because most of the information we need to transmit does not lend itself to story form. The one exception might be literature surveys used in academic dissertations, which often take the form of stories of how a certain process or piece of knowledge came to be known. In these cases the narrative starts with a time, a place, with characters being the researchers, and the plot being the mysteries of the problems. Another narrative might be the telling of how a final design was arrived at. For example, "We first tried mild steel, but discovered that the acidic environment was too harsh. We then decided to use stainless steel."

1.1.3 Description

Descriptions convey information through the five senses: sight, hearing, taste, touch, and smell. Descriptions can be part of either expository writing or narrative. Analogies and metaphors are often used in descriptions. Consider how the following sentence relies on our senses for communicating information:

The aeration basin at the wastewater treatment plant foamed over, producing billowing masses of acrid foam that were picked up by the wind and carried hundreds of yards downwind like huge, dirty clouds.

The analogy to clouds is useful and describing the foam as dirty and acrid (sight and smell) invokes a negative image, just as intended.

Descriptions can be greatly enhanced by comparisons to familiar objects. Since we visualize in five ways – size, shape, color, texture and position – a wide range of comparisons is available.[1]

For example, the size of an object can be compared to a coin (e.g. a penny), a typewriter, a car, or a tennis court. The shape of an object can be described geometrically as circular, square, cylindrical, L-shaped, or tubular. Often the shape can be related to known shapes such as a pencil, or a table, or a spider. Texture can be described as smooth, rough, coarse or glazed, while position can describe the object in relation to other known objects, such as behind the chair or parallel to the road. The objective in descriptive writing is to use as many of these comparisons as necessary in order to convey the most accurate representation of the object in words.

In some instances, a seemingly accurate description is insufficient if unwarranted assumptions are made about prior knowledge or if the words have several meanings. For example, an Englishman might describe the game of cricket in this manner:

> You have two sides; one out in the field and one in. Each man that's in the side that's in goes out and when he's out he comes in and the next man goes in until he's out. When they are all out the side that's out comes in and the side that's been in goes out and tries to get those coming in out. Sometimes you get men still in and not out. When both sides have been in and out including the not outs, that's the end of the game![2]

Even thought this is an accurate description of the game of cricket, it is very difficult to follow because simple words have different meanings.

1.1.4 Persuasion

Persuasion is the act of convincing the reader that your view is the correct one. Persuasive writing tries to change the reader's mind and uses analysis, data, and other factual material as tools to achieve this change.

In engineering, technical decisions are sometimes not open to argument. (e.g. This span requires a 10WF30 steel beam.) Engineering problems are open to many solutions (e.g. should the bridge be of concrete or steel?). Engineers who develop what they believe would be the most appropriate solution for a given problem must convince their colleagues and clients that their idea is indeed the best solution.

There are two basic techniques for organizing a persuasive argument:

- State the conclusion first, then fill in the details.
- Begin by getting the reader to agree with the details, one at a time, and then present the conclusion.

In the first instance, the conclusion is stated at the outset, followed by filling in the argument. In the second instance, the argument is built up in a stepwise way so that the reader agrees with every step and cannot then fail to agree with the conclusion.

The placement of the concluding statement either first or last takes advantage of two important rhetorical techniques – *primacy* and *recency*. Audiences remember best the first sentence of each presentation since this is their first idea of what is to be presented (primacy), and the last sentence, since this is the most recent in their memory (recency).

For example, suppose the engineer is trying to convince the management to enter into a new product line, to manufacture and sell yo-yos. The engineer could start the presentation by saying, "We should start a new product line, manufacturing and selling yo-yos." This forthright statement would get everyone's attention, and the engineer would then follow with the details of the argument.

Conversely, the engineer could begin by saying, "Our company wants to make a profit. Our competition is making a profit by selling yo-yos. We could make better yo-yos than anyone else...," concluding with the statement, "We should start a new product line, manufacturing and selling yo-yos." Placing the conclusion at the end also has the advantage of producing the least opposition since it starts with facts with which nobody disagrees and ends with a conclusion based on these facts.

Persuasive writing is, however, subject to fallacies that can detract from the force of an argument. Here are a few such fallacies to avoid:

Implied assumptions include those made about the level of knowledge the reader has about the material. For example, proving a theorem using calculus is not very persuasive if the reader is not familiar with calculus. Sometimes implied assumptions are intentionally misleading. Political campaigns are notorious for this technique. The fact that a political candidate is pictured riding a horse on his ranch, wearing a red checkered wool shirt and jeans, does not necessarily mean that he is sympathetic to environmental concerns.

Begging the question occurs in cases where proof is necessary but the conclusion is simply assumed. "When are you going to stop over-billing your clients?" is a classic example of begging the question. The question being begged for is "Are you over-billing your clients?"

The **lack of cause and effect** is perhaps the most insidious fallacy that bombards us daily. The implied cause may have no bearing on the effect, or there may be many other possibilities. For example, the statement, "I didn't get an A in this course because the professor doesn't like women engineering majors" implies that the professor does not give A grades to women engineering majors. A misreading of cause and effect often hides other problems such as poor class attendance.

Another potential problem in argumentation is the **non sequitur.** Two assertions might be placed next to each other, but this does not mean that the second necessarily follows from the first. (*Non sequitur* is Latin for "it does not follow.") "Although the moon is smaller than the earth, it is not as far away as the sun," is an example of a *non sequitur*.

A most grievous sin for engineering writers is the **misuse of numbers.** The inappropriate reporting of statistics can be especially galling. For example, in one notorious misuse of statistics, an opponent of nuclear power plants reported that, in an Arizona town where a nuclear power plant had started operation, the incidence of leukemia the following year increased by 100%. That's pretty disturbing unless one also knows that there was only one case of leukemia the first year and two the second. In this situation, bad statistics are used to suggest causality.

One of the best ways to look silly in real-life engineering is to not understand the concept of significant figures. A memo to the boss stating, "I estimate that it will cost $36,389.47 to convert the production line," will make the writer look foolish, even though all the calculations were made, and the numbers added up to exactly $36,389.47. The estimate can't be that accurate, and $37,000 is OK.

There is more guidance on the importance of numbers and their use in engineering communication in Chapter 2.

1.2 FORMATS

The structure of a written work is its format. A number of common formats are used in engineering and the structure of each has become standard practice. Just as units and dimensions have shared meaning in engineering so do formats used in written communications.

The common formats discussed below are the informal note, the memorandum, the letter, the resume, the progress or interim report, and the final report.

1.2.1 Informal Notes

Although informal notes are a highly useful and expedient method of communication, there are dangers with informal notes. In an informal note the author might be careless in what is said. Matters of a sensitive nature are best handled orally and not in written notes. E-mails are stored forever and can and often are retrieved. The admonition to not write anything that would embarrass you later is especially true for informal e-mail. Because of the lack of privacy in e-mail, pretend that anything you send over e-mail is also tacked on your office door.

1.2.2 Memoranda (or more informally, "memos")

Memos differ from notes in that copies are always retained. All memos are internal to the organization, but care should be taken with their content in order to avoid future embarrassment and even possible litigation.

The structure of a memo is straightforward:

TO: _____

FROM: _____

DATE: _____

ABOUT: _____

followed by the text. Memos are not signed, as letters or even informal notes are, although some writers will initial their name in the "from" heading to indicate their personal approval of the memo.

If the memo is to more than two or three people the custom is to write, "To distribution list" after the "TO:" and then list the persons receiving the memo at the bottom of the page.

As a rule of thumb, **no memo should be longer than one page**. If what needs saying requires more space, write a full report and include a cover memo summarizing the main points.

Here is a typical memorandum that might be used in an engineering consulting firm.

To: Al Einstein, Project Engineer
From: Siggie Freud, Vice President
Date: June 13, 2007
About: Cost overruns on the Frog Pond job

I have been monitoring the cost on the Frog Pond wastewater treatment plant project (our reference OD-450-99) and find that the present costs exceed the budgeted allowances. Of greatest concern is the time charged to this job by the laboratory. Perhaps we should review exactly what they should be doing and see if we can get the cost under control.

cc: Jill Avery, Accounting

1.2.3 Letters

The main difference between a memorandum and a letter is that a letter is usually sent outside the organization and as such represents the organization to the world. If the letter is poorly written, it implies that the whole outfit is a pretty sorry place.

One good way of writing a letter is to first write one short sentence summarizing the purpose of the letter. For example, "I want a job" is a perfectly good reason to write the letter, and the letter should simply state this wish in a formal and convincing manner.

State the purpose of the letter in the first paragraph. Don't allow the reader to wander through the letter looking for the purpose. Be positive, considerate, natural, specific, honest, and most of all, concise. Sum up your purpose in a short last sentence. Remember the idea of the recency position.

The format of letters is fairly standard world-wide. European business correspondence often includes a file or project number in the heading, and it is good manners to refer to that number when responding to a letter.

As a general rule, you should never use company or agency letterhead if the letter is not part of your job. Writing a letter to a credit card company or a letter to the editor of a newspaper is not an appropriate use of company stationery.

Below are two sample letters, one to a potential employer (on plain paper) and one to a client (on engineering company letterhead.) Note that the first letter is addressed to "Dear Sir/Madam:" because the name of the person is not known. If the firm is a small one, this is quite appropriate. If it is a larger firm, get the name of the human resources (HR) person off the web and address it directly to him or her.

February 13, 2007

Campus Box 8733
Bucknell University
Lewisburg PA 17837

Kermit & Associates
13 Croakly Drive
Frog Pond, NC 05055

Dear Sir/Madam:

I will be graduating this May from Bucknell University with a BSE in Civil Engineering. From your web site I see that your firm is engaged in the type of engineering that interests me – design of water and wastewater treatment plants – and I would like to visit your office and speak to you about potential employment. Enclosed is a recent resume summarizing my experience and career objectives. I would be pleased to provide you references as needed.

Thank you for your consideration, and I look forward to hearing from you.

Sincerely yours,

Tad Pole

Tad Pole

Kermit and Associates
Consultants
13 Croakly Drive, Frog Pond, NC 05055
919-967-4654

February 13, 2008

Lily Pad
Muck Pumps, Inc.
400 Slurry Road
Schmutzdekke, NJ 12345

Dear Lily:

Pursuant to our telephone conversation of February 12, 2008, we would like to invite you to submit your recommendations for the primary sludge pumps to be installed at the Frog Pond wastewater treatment plant. The spec sheet is attached.

Given our time constraints, we need the recommendations by March 5, 2008. My understanding is that you will be able to get them to us by this date.

If you have any questions, please do not hesitate to ask.

Sincerely,

Tad

Tad Pole
Project Engineer

cc: J. J. Kermit

1.2.4 Resumes

The word describing a summary of one's qualifications is actually spelled "résumé" and is borrowed from French. Typically the accent marks over the letter e's are dropped for English spelling.

A resume is not a summary of your life, or an autobiography, but a statement of your qualifications and aspirations for a specific job or career. Include only information that would be useful to a future employer, including some items that show what you do in your spare time and what other interests you have. You need to write your resume with a specific objective in mind, which includes a description of the position you are seeking and an argument as to why you are the very best person in the world for this position.

Your resume should be laid out in a logical sequence of information.

Contact Information. Your name, addresses, telephone, e-mail address, your personal web page, etc.

Objective. State why you are sending the resume. What kind of job do you want and what kind of career do you envision for yourself. The more information you can provide here, the better is the chance that the human resources people who first read your resume will direct it to the appropriate department or office.

Education. We are assuming here that you are going to graduate with a degree in engineering. This is what you have to sell, and other achievements such as being the valedictorian in your high school class are irrelevant. Include, therefore, only your higher educational background. List the college or university and the degree attained. If it is pending, give the date expected.

Work experience. Many firms like to see some significant work experience. Having been a camp counselor, for example, is not very useful, but it should nevertheless be listed. You don't want to leave gaps in your timeline. If you list nothing for a summer, the resume reader might wonder what it is you did for an entire summer.

Relevant skills. In today's computer-driven world, include your skills in such software applications as Autocad. Often established engineering firms are keen to hire young engineers just to get the latest computer expertise.

Interests, hobbies, or other information. It is unlikely that this will get you a job, but it might be a way of "breaking the ice" at a job interview. If you like backpacking, and you discover that the interviewer is also an avid hiker, you might spend some time discussing your travels on the Appalachian Trail.

Some people will include references on the resume, but this is not recommended. You should always talk to your recommender before you use him or her as a reference, and it is much better therefore to include your reference names and addresses in the cover letter. This allows your resume to be specific to the job you are seeking. Do not put this lame statement at the bottom of your resume:

"References provided on request."

Of **course** you will provide references. This is understood.

Here is a typical resume for a graduating engineering student:

18

RESUME

Tad Pole

Contact information:

Mailing address........Campus Box 8733
Bucknell University
Lewisburg PA 17837

Telephone...............570-577-1234 (dorm)
404-567-3456 (cell)

e-mail...................pole@bucknell.edu

Career objectives: Design and operation of water and wastewater treatment plants, hazardous waste management, general environmental engineering.

Education: BS in Civil Engineering, Bucknell University (expected May 2007). Grade point average: 3.2/4.0.

Work Experience:
Summer 2004: Laborer, town of New London, NH. Painting fire hydrants and general maintenance.
Summer 2005: Counselor at Camp Tionesta, Boy Scouts of America, teaching swimming and boating
Summer 2006: Research Assistant, Department of Civil and Environmental Engineering, Bucknell University. Conducting experiments at local landfill on biogas formation from food wastes, under the direction of Professor Thomas DiStefano.

Computer Skills: Proficient in Autocad, Mathlab, spreadsheets, and computer-aided design packages.

Extracurricular Activities: Bucknell Wind Symphony, intramural basketball, backpacking, Habitat for Humanity.

1.2.5 Progress or Interim Reports

Often engineer need to submit periodic assessments of progress, or drafts of preliminary conclusions for discussion with the client. These reports can be informal only in their appearance. The substance of the report must be as carefully considered as the final report because no matter how large the letters are that proclaim

DRAFT

all over each page, the report may well find its way into the hands of readers who might at some later date challenge or embarrass the author and/or the firm. If possible, the illustrations used in a draft report should also be of a sufficiently high quality to avoid misunderstanding or embarrassment in the future. The format of the interim report should be essentially the same as the final report and the rules concerning format for the final report apply equally for the progress report.

1.2.6 Final Reports

The final report is to engineering firms what the symphony is to serious composers – the culmination of long study and the concentration of the best professional efforts possible. The final report is what the engineer presents to the client, and this document becomes the instrument by which the engineer is judged.

The format for final reports is not rigid because the type of engineering done and the client's requirements can vary widely. The main point to consider is reader expectation. Who is the audience and what are the readers looking for? Do they expect to find a letter of transmittal as the first page? Do they expect a summary, a background chapter, a glossary? If this is a report to the president of a private corporation, a glossary is probably unnecessary. On the other hand, if this is a report on siting a solid waste landfill and it may be read by thousands of people who do not know the engineering terms, a glossary might be helpful.

When the client is a public entity and the final report becomes a matter of public record, it is good practice to organize the final report along these lines:

1. **Title page**. Spend time on the title and make it a good one. The title is the first part of the report read, and there is no second chance for a first impression. Put first the most important words in you title (the primacy position). A title that begins: "A Summary Report on the Evaluation of the Several Projects to...." is pretty much worthless. One of the first two or three words has to be a key word that describes the subject of the report.

2. **Letter of transmittal**, summarizing the reason for the report and thanking the client for the opportunity to work on the problem. A paragraph describing the most important conclusion may be added.

3. **Acknowledgments**, including the personnel employed by the client who helped you with the study.

4. **Table of contents**.

5. **List of figures** and **list of tables**.

6. **Summary** of the study, in a few pages, depending on the size of the report. The summary must be able to stand alone, because it may be copied and distributed without the rest of the report. In technical journals this is called an **Abstract** and its role is the same, to provide a short stand-alone statement of the essential contents of the paper.

7. **Background material**, ending with the objective of the report.

8. **Text** or body of the report.

9. **Conclusions**. These belong in the report and can not be extracted. Conclusions therefore can refer to the body of the report and have references. (The conclusions in the letter of transmittal must, of course, be exactly the same as in the body of the report. The letter of transmittal is a separate part of the document and can stand by itself. The conclusions at the conclusion of the report cannot stand by themselves and must be supported by the rest of the report.)

10. **Appendices** include supporting material that is not necessary for understanding the report but may be useful for subsequent studies.

To summarize: Anticipate what the reader expects, and write to that expectation. Surprises are not welcomed by readers of technical reports.

Recommended templates for the front pages of a final report, including a sample letter of transmittal, are shown on the next few pages.

FINAL REPORT
Frog Pond
Wastewater Treatment
Feasibility Study

July 2009

Project Team

Frog Pond Regional Commission
Town of Frog Pond
Mecklenburg County
Gastonia County
Monroe County

Kermit and Associates, Consultants

Kermit and Associates
Consultants
13 Croakly Drive, Frog Pond, NC 05055
919-967-4654

July 3, 2009

Mr. I. M. A. Toad
Planning and Programming
Frog Pond Regional Commission
200 Lily Pond Street
Frog Pond, NC 27516

Subject: Frog Pond Wastewater Treatment Feasibility
Study

Dear Mr. Toad:

Kermit and Associates is pleased to submit this final
report summarizing work on the Frog Pond Wastewater
Treatment Feasibility Study. Submission of the report
completes our assignment under the Contract for
Professional Services dated June 13, 1996.

Our work on the study involved a comprehensive analysis of local and regional operations for the management of wastewater discharges to the Chattahoochee River from the Town of Frog Pond and the counties of Mecklenburg, Gastonia and Monroe through the year 2040. Over the next 50 years, wastewater flows in the study area are expected to increase from 298 million gallons per day (mgd) to 644 mgd. To determine the best methods for accommodating this increase in flow we evaluated conventional treatment and discharge options as well as alternative technologies such as water conservation, land treatment and disposal, and water reclamation and reuse. Improvements to wastewater conveyance systems also were considered, including construction of rock tunnels to transport regional flows.

Based on the results of the economic and non-economic evaluation of alternatives, we recommend that the Frog Pond Regional Commission adopt a plan for the study area that relies on expansion and upgrading of the local conveyance and treatment systems by the individual project participants. The recommended alterative has significant non-economic advantages. It provides flexibility for adaptability to future changes in regulatory and institutional policy, and it is compatible with water reclamation and reuse as a future water supply supplement for the metropolitan Frog Pond area.

The total cost of the 50-year capital improvement program is approximately $1.7 billion in 2009 dollars. Over half of this amount, or about $0.9 billion of construction will be required in the next ten years, assuming that growth rates of the early 2000s will continue.

We have enjoyed working with the Frog Pond Regional Commission on the project and appreciate the excellent support and cooperation extended to us by you and your staff as well as the staffs of the five project participants.

Very truly yours,
KERMIT AND ASSOCIATES

J.J. Kermit

J. J. Kermit
President

Enclosure: Final Report (50 copies)

Technical Subcommittee Members

Frog Pond Regional Commission

William Clarkson
Peter Lalor
James Nissen

Town of Frog Pond

Kelly Derr
Bill Hanson

Mecklenburg County
Steve McCullers
David Word

Gastonia County

Ted Jett
Jack Dozier

Monroe County

Lisa Miller
Stig Regli

Kermit and Associates Project Staff

ENGINEERING

J. J. Kermit, Project Director
Frank McAlister, Project Manager

Dale Webster
Hal Davis
Lauren Bartlett
Janice Bartlett
Barbara Simpson
Gary McConnell
Raycharn Liou
Tsau-don Trai
Banu Ormeci

Aysen Ucuncu
Tom Ramsey
Tai-Yi Su
Laura Steinberg
Bob Rooks
Gary Anderson
Dilek Sanin
Li-chung Chuang

REPORT PREPARATION

Gerry Hartman
Paul Spence
Jerry O'Brien

Sue Hurst
Mich Dupré
Robert Palimeno

GRAPHICS AND ILLUSTRATION

Bill Martello
Brenda Harrison

Dan Ward
Mark Menetrez

Subconsultants and Contractors

Che-Jen Lin Earth Systems, Ltd.
Athens, Georgia

David Cobb Sludge Shovelers, Inc.
Boston, Massachusetts

Richard Henrikson and Bill Worrell
Somewhere, California

Jody Eimers Procrastination, Ltd.
Raleigh, North Carolina

Mike Uhl Reprographics
Morgantown, West Virginia

Table of Contents

1.5 A PROCESS OF WRITING

The first logical step in writing is to define for yourself the subject and purpose of the communication. One suggestion is to start with the words "I want to tell you that..." and then cross them out when the letter or report is finished. After defining the subject and purpose, you might even realize that it isn't worth communicating (the ultimate brevity!)

Next, define the audience. Who will be reading this and how do they relate to the subject and purpose? Answers to these questions will define the most appropriate type of communication. Defining the material to be communicated and the effect it is to have sets both the beginning and the end. All that's necessary now is to fill in the middle.

There is no best way to fill in the middle. Writing is a highly individualistic act, and while some authors labor for days over a single paragraph, others write entire books in a week. There is no one rigid writing process. Nevertheless, here are some pointers that may be useful for beginning technical writers:

1. Prepare an outline. Try to get the central ideas into the outline and develop a logic of thought. Do the ideas flow and form a complete thought? Some writers purposely use a tiny sheet of paper for the initial outline to prevent them from spending a lot of time refining it. Others will use word-processing and cut-and-paste for organizing the outline. There is also software available for making outlines and this can be useful if the logic of the report or paper is already clear.

2. Sketch all the graphs and other illustrations. It is always easier to write when the drawings are spread out in front of you, just as it is easier to give a talk using slides.

3. Write. Move it! Don't wait for the proper construction. Don't search for just the right word. Get everything into the word processor or on paper. If the piece is lengthy, don't start at the beginning. Start writing the easiest part first; then move to the more difficult sections. In project reports, for example, the procedures and methods should be the simplest to write, so start with those. Write the introduction, which is often the most difficult section, last.

4. Revise in a rough way. Look for missing ideas, poor transitions, or convoluted construction.

5. Cool it. Try to put it aside for as long as possible; two weeks is the minimum for most writers, but in a pinch, even a few hours can help.

6. If possible, and if appropriate, have others read the piece and make suggestions for improvement. Sometimes the most obvious mistakes can slip by you until someone else points them out.

7. Edit as a critical reader. With sufficient cooling time the material should read like someone else's writing. This is the chance to be especially critical of your own writing. Obviously we seldom have enough time to write, edit, cool, and edit. The boss wants the report by quitting time and it has to be done – and done well. But the more time you can spend on a piece of writing, especially with large cooling periods interspersed, the better will be the product.

32

REFERENCES

1. Houp, K. W. and T. E. Pearsall, *Reporting Technical Information,* Macmillan Pub. Co., New York, 1984.
2. Anonymous, found on an old tea towel in St. Albans, England, 1983.

PROBLEMS

1-1 Write a one page memorandum, addressed to your academic advisor, outlining your professional plans after graduation.

1-2 Write a one page memorandum, addressed to the instructor of this course, describing the worst case of cheating you have either seen or been aware of on campus. Use the narrative mode of writing.

1-3 Write a one page memorandum, addressed to your great grandfather and grandmother, thanking them for your life. Describe what you have done and will do to make them proud of you.

1-4 Write a one page letter addressed to the editor of your school newspaper arguing that Ultimate Frisbee should be a recognized NCAA interscholastic sport (with scholarships, of course). (If you are unfamiliar with Ultimate Frisbee, choose some other "sport" such as chess or Spider Solitaire.) Make sure the letter is in a proper format.

1-5 Write one short (!) paragraph in the *expository* mode. Choose whatever topic interests you. Then write three more short paragraphs in the *narrative, descriptive,* and *persuasive* modes. Be prepared to discuss in class how these paragraphs differ from each other.

1-6 Using the first page of your college newspaper, look for the use of *expository, narrative, descriptive,* and *persuasive* writing. Mark these, bring the paper to class, and be prepared to discuss why you chose these examples.

1-7 Using your college newspaper, choose one article and study its structure with regard to topical arrangement. What is the structure the writer was striving for? Should he or she have used headings? Would there have been a better way to present the material to make it more readable?

1-8 Offer five *comparisons.* They can be humorous, but in good taste. For example, you might say "The professor smiled like a Cheshire cat."

1-9 Offer five *metaphors.* For example: "The professor's question was a dagger aimed at my heart."

1-10 Choose any one of hundreds of commercials on television and discuss how the argument is flawed due to the lack of *cause and effect.* That is, what is the audience asked to believe that cannot really be rationally supported?

CHAPTER 2

COMPOSITION AND STYLE IN ENGINEERING WRITING

Clear and distinctive writing is not a natural gift, any more than structural dynamics or biological growth kinetics is a gift. It is a tool learned by doing, being corrected (edited), and redoing.

Good writing is not a new idea, of course. In 1733 Benjamin Franklin described skilled writing in this way:

> Good writing should proceed regularly from things known to things unknown, distinctly and clearly without confusion. The words used should be the most expressive that the language affords, provided that they are the most generally understood. Nothing should be expressed in two words that can be as well expressed in one; that is, no synonyms should be used, or very rarely, but the whole should be as short as possible, consistent with clearness; the words should be so placed as to be agreeable to the ear in reading; summarily it should be smooth, clear, and short, for the contrary qualities are displeasing.[1]

An essential element of good style is following rules of grammar and syntax, but this is not sufficient. Good style also means writing "with style," a blatantly circular definition. What is meant by writing "with style," or writing so that it is "agreeable to the ear," and how do we recognize such writing?

As Strunk and White[2] suggest, one means of testing the quality of any writing is to try to improve it by rewriting. For example, famous sentences and expressions defy such rewriting, since their "style" is what makes them so memorable. Strunk and White illustrate this idea by asking for a rewrite of the famous sentence by Thomas Paine:

These are the times that try men's souls.

Some variations might be;

Times like these try men's souls.
How trying it is to live in these times!
These are trying times for men's souls.
Soulwise, these are trying times.

The above variations, while grammatically correct, lack the memorable style of the original.

2.1 COMPOSITION

Composition is the nuts and bolts of style. The elements of composition are not intuitive and simply have to be learned. We begin by considering the use of words, sentences, and paragraphs in engineering writing.

2.1.1 Words, sentences, and paragraphs[*]

Use simple words.

BAD EXAMPLE: Plan A *minimizes deleterious environmental impacts.*
REVISION: Plan A causes the *least environmental damage.*

---

[*] Some of these examples are borrowed from a short pamphlet written by James G. Smith and subsequently published as Smith, J. G., and P. A. Vesilind, *Report Writing for Environmental Engineers and Scientists*, Lakeshore Press, Woodsville, NH, 1996. Used with permission.

37

BAD EXAMPLE: *Obsolescent* equipment *induces excessive energy consumption.*
REVISION: *Old* equipment *wastes energy.*

Use the correct word. Mark Twain observed that, "The difference between the right word and the almost right word is the difference between lightning and the lightning bug."

BAD EXAMPLE: Odors were *observed* near the plant. (The writer had good eyes?)
REVISION: Odors were *detected* near the plant.

BAD EXAMPLE: About 90 percent of the population *is connected* to the sewerage system. (Sounds painful!)
REVISION: The sewerage system *serves* about 90 percent of the population.

Following is a list of a few commonly misused words:

imply: to suggest
simple: uncomplicated
method: procedure

summary: a condensed version
assure: to declare positively
proceed: to move forward
forward: toward a point in time or place ahead
preparatory: designed to make ready

infer: to conclude
simplistic: oversimplified
methodology: principles applied to a branch of knowledge.
summation: a closing argument of the text
ensure: to make certain
precede: to go before
foreword: an opening statement in a report or book
prefatory: preliminary

fortunate: lucky

farther: more in distance

that: defining, restrictive

fortuitous: happens by chance

further: more in time or quantity

which: modifying, non-restrictive

GOOD EXAMPLE: The motor that failed needs to be replaced

GOOD EXAMPLE: Motors which fail need to be replaced.

disinterested: impartial

effect (n): result

effect (v): to bring about

uninterested: not interested

affect (v): to influence

EXAMPLE: The plant failure *affected* (v) the decision. The *effect* (n) of the decision was to change the engineering firm, *effecting* (v) the collapse of the firm. (The plant failure *influenced* the decision. The *result* of the decision was to change engineering firms, *bringing about* the collapse of the firm.)

Do not use nouns as verbs. This admonition has less weight today than in years past. Technology, especially computer technology, has introduced many new structures into the language, and now the use of many nouns as verbs is accepted practice. For example, one now can *access information*, even though *access* is a noun, and *impact* is now listed in the dictionary as both a noun and a verb. This development should not, however, open the barn door to the further deterioration of our language. Avoid using nouns as verbs unless you have an overwhelming reason for doing so.

Use specific rather than general words.

BAD EXAMPLE: *Most* of the site is contaminated.
REVISION: Over *75 percent* of the site is contaminated.

BAD EXAMPLE: The chlorinator failed *several* times.
REVISION: The chlorinator failed *five* times.

BAD EXAMPLE: How does that *impact* you?
REVISION: Does that *impress* you?

Note that *impact* is a hollow word. Perhaps the writer meant to say "How do you feel about that?" or perhaps the intended meaning was "How does this affect your project schedule?" or any number of alternatives. Specific is always better than vague.

Use repetition for clarity. Do not confuse a thought or issue by avoiding word repetition when the repetition increases clarity.

BAD EXAMPLE: We applied carbon at a rate of 10 pounds per day, then at 20.
REVISION: We applied carbon at a rate of 10 pounds per day, then at 20 pounds per day.

Avoid using strings of adjectives.

BAD EXAMPLE: The system employs standby, gas-engine-driven, hollow-shaft, high-head, deep-well turbine pumps.
REVISION: Standby well pumps are gas-engine driven.

Place modifiers as close as possible to the words they are intended to modify.

BAD EXAMPLE: He *only emptied* three barrels.
REVISION: He *emptied only* three barrels.

Note how changing the position of the two words completely changes the meaning of the sentence.

BAD EXAMPLE: Wind-blown dust, flies, and rodents could transport bacteria from the area of the overflow. (In this position, *wind-blown* modifies all three nouns, implying that there are wind-blown rodents.)
REVISION: Flies, rodents, and wind-blown dust could transport bacteria from the area of the overflow.

Use prepositional phrases sparingly. They often require unnecessary words and might cause confusion.

BAD EXAMPLE	REVISION
result in	cause
on the order of	about
in the event that	if
at the present time	now

BAD EXAMPLE: The project includes *the provision for* a new pump.
REVISION: The project includes a new pump.

BAD EXAMPLE: Gravity feed will resume *in the event that* the pump fails.
REVISION: Gravity feed will resume *if* the pump fails.

BAD EXAMPLE: They are *in the process of* revising the report.
REVISION: They are revising the report.

There are two ways to reduce the number of prepositions. The first way is to rewrite the sentence using the object of the preposition as an adjective.

BAD EXAMPLE: The *director of the laboratory* was ill.
REVISION: The *laboratory director* was ill.

The second way to reduce prepositions is to reuse the object of the preposition in its possessive form.

BAD EXAMPLE: The impeller *of the pump* is worn out.
REVISION: The *pump's impeller* is worn out.

When the text refers to figures or tables, use the proper preposition. Items are *on figures*, or *in tables*.

BAD EXAMPLE: The likelihood *of becoming* a better writer *by reading from this chapter on writing skills* is *of slight probability*.

Avoid using gerunds as sentence subjects. (Gerunds are verbs ending in *-ing* used as nouns.)

BAD EXAMPLE: *Trapping* of the heavier sediments occurs in the lake.
REVISION: The lake *traps* the heavier sediments.

Use appropriate tense. Because present tense is the simplest tense and thus least prone to confusion, use present tense in all cases except where the action has clearly occurred in the past or will occur in the future. Note that a report or a book (such as the one you are reading) exists, so Chapter 1 is referred to in the present tense, as is Chapter 6. If you are writing Chapter 6 and refer to something in Chapter 1, use present tense because whatever you are referring to already exists at the time you are reading it. Similarly, if you are writing Chapter 1 and refer to something in Chapter 6, use present tense because once again when the reader is reading Chapter 1, Chapter 6 already exists.

BAD EXAMPLE: In Alternative 2, the plant *was* abandoned.
REVISION: In Alternative 2, the plant *is* abandoned.

BAD EXAMPLE: The calculations *showed* that the flow *was* inadequate.
REVISION: The calculations *show* that the flow *is* inadequate.

BAD EXAMPLE: Public speaking *will be* discussed in Chapter 3.
REVISION: Public speaking *is* discussed in Chapter 3.

When describing work performed during a project, use the past tense.

EXAMPLE: The board *considered* three alternatives.

When discussing the structure of alternatives, use the future conditional verb tense.

EXAMPLE: If the city continues its eastern expansion, a riverside location *should be* selected.

When discussing your recommendation, use the present tense, although some professionals prefer the future tense.

EXAMPLE: Plant maintenance *will* require three operators full time.

- or

EXAMPLE: Plant maintenance *requires* three operators full time.

Singular and plural of words derived from Latin must be used correctly. For example, "data" and "strata" are both plural, while "datum" and "stratum" are singular. Other Latin problem words to look out for are: "memoranda/memorandum", "alumni/alumnus (masculine) and alumnae/alumna (feminine)."

BAD EXAMPLE: The data *indicates* that . . .
(plural noun, singular verb)
REVISION: The data *indicate* . . .

Be positive. Avoid negative expressions.

BAD EXAMPLE: We do not believe that the interceptor is adequate.
REVISION: A new interceptor is required.

Take great care with gender-specific pronouns. If the gender of the person is unknown, the pronoun might be *she or he*, *he/she* or *him/her* or *s/he*. An alternative is to use the grammatically incorrect *they*. Most of the time gender-specific terms can be eliminated in engineering writing without loss in either style or accuracy.

BAD EXAMPLE: Three *men* are required . . .
REVISION: Three *operators* are required . .

BAD EXAMPLE: The project manager must have tight control on *her* budget.
REVISION: The project manager must have tight

control on *the* budget.

BAD EXAMPLE: Every project manager should monitor *his* project carefully.
REVISION: Every project manager should monitor *his/her* project carefully.

or use a plural subject,

REVISION: Project managers should monitor *their* projects carefully.

or change the pronoun to plural (grammatically incorrect)

REVISION: Every project manager should monitor *their* project carefully.

Avoid uncertainty. Limit the use of *would, could, may,* or *might* to truly uncertain situations.

BAD EXAMPLE: The treatment process *may be* the best available.
REVISION: This *is* the best treatment process available.

Use strong nouns and verbs. Some abstract nouns are: *basis, case, condition, effect, facility, factor, method, nature, reference, situation, system,* (and, blandest of all) *thing.* Be specific in technical writing. For example, the sentence "This engineer was finishing the work while communicating with others," is not nearly as informative as saying "Alice was editing the Frog Pond report and talking on the telephone to Jess in the duplicating department."

BAD EXAMPLE: The water must be cold.
REVISION: The water temperature must be below 8°C.

BAD EXAMPLE: The handle must not be too high off the floor.
REVISION: The handle must not be higher than 29 inches from the floor.

The same rules apply to verbs. Consider how weak these verbs are: *is based, is considered, is facilitated, is made, is done, is utilized.*

BAD EXAMPLE: The lamp broke.
REVISION: The lamp shattered.

In either case, we know that the lamp is no longer whole, but the second sentence tells us a great deal more about what happened to it.

Don't use empty words and phrases. Empty language in technical writing should be avoided. Some particularly obnoxious phrases that often occur in business letters are:

> with your permission
> herewith enclosed
> inasmuch as
> in view of
> with regard to
> with reference to

Avoid pompous words. The three favorite engineering pomposities are *utilize, terminate,* and *viable.* Simple straightforward words will do, such as *use, end*, and *living.*

Don't get trapped by hidden antecedents. Words like "it" and "this" can cause problems. The formal rule is that "it" refers to the immediately preceding noun, but most people do not remember such niceties. For example: "The class enjoyed the field trip to the wastewater treatment plant although it smelled bad," is a perfectly good sentence ("it" refers to the preceding noun, "plant") but the sentence sounds funny because we aren't sure if the word *it* refers to the plant, to the field trip, or to the class.

> BAD EXAMPLE: Power loss to the control panel caused premature closure of the valve and sudden stoppage of the pump. The manufacturer inspected *it* the following morning.
> REVISION: Replace *it* with: *control panel, valve, pump* or *system*, depending on the true antecedent.

Avoid euphemisms. "Bathroom" and "pass away" are two favorites. Use "toilet" and "die," if this is what you mean.

Don't create new words. A fascinating phenomenon in engineering writing is the use of words ending in "-ize" and "-wise." Sometimes they are necessary, but try to avoid creating new words.

> BAD EXAMPLE: Odorwise, he minimized public opposition.

Don't confuse "should" with "shall." This is an important distinction when writing legal documents such as engineering specifications. The word *should* leaves it up to the contractor to do or not to do something, whatever is to the contractor's advantage. The word *shall* on the other hand makes a specification mandatory. Thus, "A handrail *shall* be constructed for the temporary stairs," means that the handrail is mandatory, while "A handrail *should* be constructed for the temporary stairs" is merely a suggestion.

Never forget that sometimes bad engineering writing not only costs money but may endanger lives. A writer has a responsibility for safety and must be aware of manufacturing defects, design problems, inadequate warnings, or unclear instructions. All products that can cause injury, even if used incorrectly, must be properly labeled, preferably with one of three words, *caution*, *warning*, and *danger*. The label *caution* means that even with normal care, harm *may* occur. *Warning* is stronger and means that without precautionary measures, harm *will* occur. Finally, *danger* announces an imminent and severe hazard to be avoided altogether.

Vary the lengths of sentences. A string of long sentences can be hard to read, while a string of short, jerky ones may require the reader to combine diverse ideas, a job that should have been done by you, the writer.

If a sentence is too long, prune it, break it in two, or divide it with a semicolon. When read aloud, a sentence should not leave the reader short of breath.

> BAD EXAMPLE: Late season withdrawals, after development of low dissolved oxygen levels, will permit release of nutrients recycled from the reservoir bottom, again only of major concern if waters are discharged directly to the stream, as could occur to satisfy the water rights permit or a possible court decree.
> REVISION: Late season withdrawals, after development of low dissolved oxygen levels, will permit release of nutrients recycled from the reservoir bottom. Such withdrawals are of major concern only if waters must be discharged directly to the stream to satisfy the water rights permit or a possible court decree.

Use the active voice. In the active voice, the subject acts; in the passive voice, the subject becomes the object of the verb's action. The passive voice makes it possible to write sentences without saying who or what is acting. One result of using passive voice is that nobody takes responsibility for the action. In addition, passive voice constructions often add prepositional phrases and make the sentences unnecessarily complex.

BAD EXAMPLE: (Passive) It was decided to change the budget. (Who did the changing?)
REVISION: (Active) *The management team* agreed to change the budget.

BAD EXAMPLE: (Passive) It was suggested that the construction schedules be revised by the project manager. (Who did the suggesting? Why is *project manager* buried in a prepositional phrase?)
REVISION: (Active) *The client* suggested that the project manager revise the construction schedule.

BAD EXAMPLE: (Passive) It was decided by the commissioner that more decisions should be made by the regional offices.
REVISION: (Active) *The commissioner* decided that the regional offices should make more decisions.

Occasionally, you must use the passive voice. In such cases, remember to put the most important words first in the sentence.

EXAMPLE: (Active) The golf ball hit the Senator. (The subject does the acting, but the emphasis is wrong.)
REVISION: (Passive) The Senator was hit by the golf ball. (A Senator is usually more important than a golf ball.)

Omit useless words. Delete useless words and phrases such as:

Too Wordy	Better
due to the fact that	because
in the process of	is (are)
during the period of	during
at this point in time	now

BAD EXAMPLE: *It can be seen* from Table 3 that an acceptable minimum flow is constantly maintained. REVISION: *Table 3 shows* that an acceptable minimum flow is constantly maintained. (Note that eliminating deadwood also pulls the sentence from passive to active voice.)

Ordinarily, writing in the active voice makes a report twenty to thirty percent shorter. Editing deadwood decreases the length further. However, shorter is not always better.

EXAMPLE: The volatile suspended solids loading at the treatment plant during wet weather . . . INAPPROPRIATE REVISION: The treatment plant wet-weather volatile suspended solids loading . . .

Watch out for *non sequiturs*. As introduced in Chapter 1, statements that do not logically follow one another are called *non sequiturs* (literally, *it does not follow*).

BAD EXAMPLE: La Playa de las Pulgas should prove attractive to light industry. Of the town's 1,292,000 people, 96,800 have moved there since 1990.

Subordinate dependent clauses to the main thought. Use words such as *because, although,* and *until* to show how the ideas relate to each other.

BAD EXAMPLE: The carbon brushes in the motor
were badly worn and the centrifuge was inoperative.
REVISION: The centrifuge was inoperative
because the carbon brushes in the motor were badly
worn.

Use punctuation to clarify emphasis. Clarify confusing
sentences with commas. Note below how the placement of
a comma changes the meaning.

BAD EXAMPLE: As discussed earlier discharges were
highly toxic.

By the placement of a comma, this sentence might one of
two meaning.

REVISION: As discussed earlier, discharges were
highly toxic.
or,
REVISION: As discussed, earlier discharges were
highly toxic.

Avoid redundancy.

EXAMPLE: The plan should be clear and
unambiguous.
REVISION: The plan should be clear.

EXAMPLE: The board stipulated these
requirements:
REVISION: The board stipulated that …

Beware of redundant phrases such as:

advance planning
other alternative
blue color
totally destroyed
work tasks
revert back
continue on
temporarily suspended
PIN number
basic fundamentals
illegal crime
most unique
past history
adequate enough
completely full

Use parallel structure in sentences. Parallel structure means that words are balanced with words, phrases with phrases, and clauses with clauses.

BAD EXAMPLE: Durable and since it costs less than the alternatives, the widget is the best choice for us.

Note that "durable" is a word while "it costs less than the alternatives" is a clause.

REVISION: The widget is the best choice for us because it is durable and costs less than the alternatives.
or,
REVISION: Durable and inexpensive, the widget is the best choice for us.

Use an effective first sentence for each new paragraph. As noted in Chapter 1, the first sentence in a paragraph is in the *primacy position* and has two functions. First, it is a transition, referring to familiar material and linking that with what is to come. Second, it introduces the material that is to be the main subject of the paragraph. In the second capacity it serves as the *topic sentence* for the paragraph. First sentences are important because readers will often skim the material and expect to see a sentence at the beginning of a paragraph that will then be elaborated in the remainder of the paragraph.

Use an effective last sentence in each paragraph. The last sentence has two purposes. First, it ties into the first sentence of the next paragraph, thus helping the reader know what to expect next. Second, the last sentence is in the *recency position* and is therefore most likely to be remembered.

> EXAMPLE: The accident evaluation team concluded that the most probable cause of the explosion was a malfunctioning catalytic reactor.

The stress is on "catalytic reactor", and this acts as a link with the first sentence in the next paragraph, which might describe why they settled on this conclusion.

> EXAMPLE CONTINUED: The catalytic reactor had been unstable for the past few weeks, and . . .

Placing the "catalytic reactor" at the end of the first sentence illustrates the principle of recency in that the last part of the sentence is the information that is most recent in the reader's memory and is therefore what is remembered best. The topic sentence should appear at or near the beginning of the paragraph, engage reader interest, and be followed by an orderly sequence of supporting information.

Conclude the paragraph with a transition into the next

paragraph or thought. A smooth transition between major points is essential to the logical progression of ideas.

Paragraphs will and should vary in length, but as a rule, they should be short to help the reader follow the text's organization. Paragraphs in engineering writing seldom exceed eight lines.

An independent statement may justify a one-sentence paragraph. In general, however, a one-sentence paragraph indicates improper consolidation of facts. Check to see if the one-sentence paragraph is just an errant point you should make elsewhere.

2.1.2 Items in series

Use a similar word structure for all items in a list.

EXAMPLE: Project financing costs include principal, interest, charges for bond sale, fees for attorneys and administrative costs. (Were there fees for administrative costs?)

REVISION: Project financing costs include principal, interest, bond sales charges, attorneys' fees, and administrative costs.

Choose a logical sequence when listing items. Sometimes it is appropriate to list the simplest items first, while at other times a chronological order might be best. In the example below, the list is by order of importance.

EXAMPLE: Domestic and industrial waste discharges, impoundments, groundwater withdrawals, irrigation return flows, and rainfall affect river quality.

For a series of items, semicolons can enhance clarity. If the series is complicated, set items apart with semicolons, not commas.

BAD EXAMPLE: Current operations include
screening, sedimentation, biological oxidation,
disinfection, final filtration and chlorine detoxification,
a high-lime process to remove phosphates, heavy
metals, organics, and viruses, and denitrification.
REVISION: Current operations include screening;
sedimentation; biological oxidation; disinfection; final
filtration and chlorine detoxification; a high-lime
process to remove phosphates, heavy metals, organics,
and viruses; and denitrification.

Note that this series follows a process flow scheme, and the
simple-to-complex structure is inappropriate in this case.

**Always place some form of punctuation between the
next-to-last and last items.**

EXAMPLE: The colors were blue, red, and green.

**Be aware that numbered, lettered, and bulleted lists do
not imply the same thing.** A numbered list implies
sequence, and this would be appropriately used for
instructions.

EXAMPLE: A typical Sousa march has these parts:

1. introduction
2. first strain
3. second strain
4. trio
5. breakup strain
6. stinger.

A lettered list implies some sequence, but is more
commonly used to imply importance, with the first item

listed under "a" being the most important and the importance diminishing down the list.

EXAMPLE: The committee consisted of

a. the mayor
b. a city council member
c. the city manager
d. the planning director.

Finally, the bulleted list implies that all of the items in the list deserve equal weight.

EXAMPLE: Intramural sports available to the students were

- basketball
- tennis
- wrestling
- flag football.

Note that the above three examples are complete sentences and require a period at the end.

2.1.3 Capitalization

Use capital letters carefully and consistently. Unnecessary or inconsistent capitalization detracts from the appearance and content of a page. Adhere faithfully to the capitalization used by the company, agency, or other organization.

EXAMPLE: The City of Twin Falls...

Generally, do not capitalize when substituting the words

city, *county*, *district*, *agency*, or *state* for the proper name of an organizational unit.

> EXAMPLE: St. Paul had such a problem. The *city's* water system...

Capitalize only the full name of national, state, county, and city agencies and organizations.

> EXAMPLES: The Seattle City Council...
> We abided by the *city* council decision to...

Do not capitalize general references to *federal* or *state*.

> EXAMPLES: The *federal government*... with *federal aid*... to the *state agency*.

Capitalize the names of geographical areas such as the South, Midwest, or Southern California.

Capitalize the names and titles of engineering studies and reports, even prior to their publication.

> EXAMPLE: *The Tucson Regional Facility Plan....*

However, a general reference to the facility or the plan does not require capitalization.

> EXAMPLE: For Tucson, *the regional facility plan....*

Capitalize titles of honor and respect that precede personal names.

> EXAMPLE: Presiding at the meeting was *Chairwoman* Jennifer Jones.

The initial reference to a person should include the official title, first name or initials, and surname. Subsequently, use only the title and surname.

EXAMPLE: The speaker was *Professor Farley P. Corey. Professor Corey* gave the Tacoma City Council a short quiz. The city council failed miserably.

Capitalize titles of high-ranking international, national, and state officials even when used in place of a personal name.

EXAMPLE: *The President*, at his weekly press conference, spoke...

Do not abbreviate or capitalize occupational titles such as *consulting engineer*, *geologist*, or *project manager* when these titles appear in sentences. Capitalize job titles when they appear after the proper name in addresses or business cards.

EXAMPLE: The geologist on the committee was Janice P. Logan.
EXAMPLE: Janice P. Logan, Geologist.

Official titles such as *professor* or *chair* can be used with a last name alone, but occupational titles can not.

EXAMPLE: According to Professor Jones, ...

BAD EXAMPLE: The award went to engineer Smith.
REVISION: The award went to engineer Roger Smith.

Do not capitalize a title when it (a) stands alone, (b) follows a personal name, or (c) is followed by a personal name in an appositive.

EXAMPLE: Mr. Aristotle Smith, *president* of the company, will speak at the meeting.
EXAMPLE: The *chair's* ruling was negative.
EXAMPLE: Who *chaired* the meeting of the Board of Directors?
EXAMPLE: The *chair* of the Board of Directors, Jennifer Jones, presided at the meeting.

When a person has a long title, place it after the personal name.

EXAMPLE: Jill Jones, executive officer of the Regional Water Quality Control Board, Central Region, …

2.1.4 Punctuation

Proper punctuation leads to sentence clarity and provides consistency in engineering and scientific documents.

Commas can and often are misused. The comma is like a yield sign in traffic. It does not mean full stop, but it does indicate a pause.

Correct use of commas includes separating a series of three or more words.

EXAMPLE: Manny, Moe, and Jack went fishing.

Commas can also be used to enclosed parenthetic expressions.

EXAMPLE: Michele, you will be glad to hear, has a new job.

Commas are typically used to separate an independent clause beginning with a conjunction.

EXAMPLE: His grades were low, but there was still a chance he could pass the course.

Commas are not used between two independent clauses.

BAD EXAMPLE: His grades were low, there was still a chance he could pass the course.

Most of the time it is proper to separate independent clauses into sentences.

REVISION: His grades were low. There was, however, a chance he could pass the course.

Semicolons are used between independent clauses when one clause explains or refers to the other.

EXAMPLE: His grades were low; there was still a chance he could pass the course.

As noted above under *Items in Series*, the semicolon can also be used to set apart a list of complex items within a sentence.

Colons are used after an anticipatory expression that directs attention to a series of items.

EXAMPLE: Engineers must furnish the following items for field assignments: flashlight, camera, insect repellent, fishing rod.
EXAMPLE: When starting the motor: 1) Read the instructions, 2) Press the START button, 3) Run for your life.

Contractions should usually not appear in engineering

60

documents. Replace contractions with the derivative words.

BAD EXAMPLE: *It's* too complicated.
REVISION: *It is* too complicated.

BAD EXAMPLE: *We're* just going to give up.
REVISION: *We are* just going to give up.

Dashes signify a sudden change, an addition of emphasis, or parenthetical material. A dash in most typewritten text consists of two hyphens with no space before, between, or after, but most computer-edited work the dash is a single line with a space before and after the line.

EXAMPLE: The pump--despite its age--does not need to be replaced.
EXAMPLE: It is a new process – and we already have it patented.

Ellipses marks indicate an omission within a quotation. If words are omitted from a sentence, the marks are three periods with no spaces.

EXAMPLE: The report showed that "… the health effects were negligible."

Hyphens are used to indicate a compound word or to break a word at the end of a line. There are no spaces before or after a hyphen. Because language usage changes over time, some compound words (especially prefixed words) that were formerly hyphenated are now written as one undivided word. For example, the word *cooperate* evolved from *co-operate* to *coöperate* to *cooperate*. Another example, one which engineers often encounter, is *decision makers*. It evolved to *decision-makers* and now it is common to see it as *decisionmakers*.

Use a hyphen to link a compound adjective when it precedes a noun.

EXAMPLE: She touched the *red-hot* coil.

Do not use a hyphen to link an adverb to an adjective when it follows a noun.

BAD EXAMPLE: The coil was *red-hot.*
REVISION: The coil was *red hot.*

Hyphenate *cost-effective* and *cost-effectiveness.*

Use a hyphen to prevent mispronunciation. For example, use a hyphen to distinguish *un-ionized* from *unionized* and *re-cover* from *recover.*

Use a hyphen to avoid doubling the vowels *a, i,* and *u,* or tripling consonants.

Use a hyphen to join a prefix to a capitalized word.

EXAMPLES: *post-Civil War, un-American.*

Hyphenate numeral-unit combinations.

EXAMPLES: *...6-foot* door ... *50-year* life ... *30- and 60-minute* intervals.

Do not use a hyphen to replace the words *to* or *through.*

BAD EXAMPLE: The project lasted March-May.
REVISION: The project ran from *May through June.*

If two words are used to modify a noun and a hyphen would be required for each, hyphenate each modifier.

62

EXAMPLE: ... *post-* and *pre-war* industries.
Observe the following rules when breaking a word at the end of a line:

- When possible, break a word on the accented syllable.
- Never divide a two-syllable word ending in *-ed,* or *-ly,* (such as *stated* and *really*).
- Avoid breaking words in a manner that leaves one or two letters at the beginning or end of a line.
- Break compound words only at existing hyphens (such as engineer-in-training).
- Avoid hyphenating words on two lines in succession; never hyphenate words on three lines in succession.
- Avoid hyphenating the last word on the first or last complete line of text in a paragraph.
- Never hyphenate the last word on a page.

Quotation Marks. Quotation marks indicate that someone is being quoted verbatim, or to identify a word that is being discussed. Other punctuation such as periods are placed inside the quotation mark.*

EXAMPLES: He said, "Let's do it." He said, "Let's go for it!" He said, "We should do this?" "But why?" he asked.

---

* Should punctuation marks such as the period go before or after the quotation mark? It depends on where you live. In Britain the period (full stop) ends the sentence and all letters and punctuation are before the period. In America, however, the common style is to place the period before the quotation mark. This practice began in the days of hot lead type when it was likely that the unsupported period would be easily damaged during the print run. So the printers here started to protect the period with quotation marks, placing the period between the last letter and the quotation marks. Logically, the period should end the sentence, and the Brits are right, but logic does not prevail in writing style.

If the quote is a long one, say more than 40 words, display it by indenting the material equally from the left and right margins. Do not use quotation marks to enclose material displayed in this manner.

Signify a quotation within a displayed quotation with quotation marks. Often the quotation is shown in slightly smaller font to indicate that it is not part of the text.

Question Marks in parentheses follow a statement that signifies uncertainty.

EXAMPLE: The wood stave pipes were replaced in 1910(?). (Note the passive voice. This is necessary if we have no idea who actually replaced the wood pipes.)

Many requests, suggestions, and commands are phrased as questions but may not require a question mark.

EXAMPLE: May I suggest that you call in advance.

2.1.5 Numbers

In most instances use words to express numbers one through ten, use numerals to express numbers over ten, and use numerals to express units of measurement and time.

EXAMPLES: ...a *two-person* review board. ...that the *17 tasks* to be performed ...ran a *6-hour* test to determine...

Use words and numerals in the same phrase when appropriate.

EXAMPLES: ...in which we propose to conduct a *4- week* study.
We recommend a *two-phase* analysis.

Use a word to express a number that begins a sentence.

EXAMPLE: *Eleven* of the 12 chapters are complete.

When two numbers appear in a phrase, use the highest value in a phrase to determine whether to use numerals or to use words.

EXAMPLES: ... *one* chance in *ten...1* chance in *10,000...*

Spell out ordinal numbers.

EXAMPLE: *Twenty-fifth* (not *25th*)

Express fractions as *1/4* or *1/2* (not ¼ or ½) unless they are part of an equation.

When two numbers come together and one is part of a compound modifier, express one of the numbers as a numeral and the other as a word. As a rule, spell out the first number unless the second would make a significantly shorter word.

EXAMPLES: ...*two 2-hour* tests...*20 two-hour* tests

Numbers expressed as numerals are made plural by adding an *s* without an apostrophe.

EXAMPLES: ... in the *1980s* ...a temperature in the *40s.*

Use numerals for all references to millions.

EXAMPLE: *$1 million, 2 million* people

Use numerals with percent calculations. Always spell out the word *percent* when used in a sentence.

Especially in the writing of specifications or other quasi-legal documents, it is sometimes useful and prudent to spell the number first in letters and then write the numeral in parentheses.

EXAMPLE: The pumping station shall have *six (6)* centrifugal pumps.

Use numerals to express units of time.

EXAMPLES: *9 am* (not *9:00 am* or *nine am*)...*8:45 pm...12 noon* (or *midnight*)

Use numerals to express units of measurement.

EXAMPLES: ...a total head of *149 feet.* At *24 mgd*, the sedimentation basins... of the top *2 feet* of sand.

Use numerals with temperature measurements.

EXAMPLE: The beer was at 54°F.

Use a hyphen between a number and a unit of measurement that form a compound modifier immediately preceding a noun. Sometimes a small modification can eliminate compound modifiers.

EXAMPLES: ... a *100-meter* tower... a tower *100 meters* high.

For clarity, use the word form of numbers preceding measured units.

BAD EXAMPLE: … the *2 1/2-inch* pipes … (this could mean pipes of 2.5 inches in diameter)
REVISION: … the *two 1/2-inch* pipes …

Use the word *foot* when a numeral and a unit of measurement form a compound modifier immediately preceding a noun.

EXAMPLE: … a *10-foot* buffer zone.

Use the word *feet* to indicate plural units when a numeral and a unit of measurement do not immediately precede and form a compound modifier of a noun.

EXAMPLES: …two paddle wheels, each *32 feet* long
… the channel is *80 feet* wide.

Express fiscal year with a slash or by the calendar year in which it ends.

EXAMPLES: Fiscal year *1990/91*, FY *1990/91,*
Fiscal year *1991*, FY *1991*

Use numerals for dates. In the United States the month comes first, followed by the day and then year. A comma is placed between the day and the year.

EXAMPLE: *December 23, 2009*

The rest of the world writes dates with the day first, followed by the month and then the year, with no punctuation. This is, of course, much clearer and more reasonable. The smallest unit, day, comes first, followed by the month, and then the largest unit, the year, at the end.

EXAMPLE: *23 December 2009*

When engineering documents are written in the United States, the local conventions should be followed. If the document is also for overseas readership, the international notations should be used.

A similar problem exists with street numbers. In the United States the house number comes first followed by the street name.

EXAMPLE: 142 Hilltop Place

But for the rest of the world, the street comes first, followed by the house number.

EXAMPLE: Hilltop Place 142

Once again, the international system seems to be more rational, since the house number is subordinate to the street name. [*]

Do not use decimal points and zeros with even sums of money unless amounts in dollars and cents are mentioned in the same paragraph.

---

[*] But curiously, international as well as United States addresses still place the person's name first, and the country of residence last. If logic governed, addresses would be written so that the largest geographical area is first, then the town, then the street, then the number, and finally the name. So my address would be

USA
New Hampshire
Bath
Hummingbird Lane 46
P. A. Vesilind

But we are stuck with convention. I wonder if I would get the letter if you wrote to me starting with USA.

68

EXAMPLES: The contract price was *$2,500.* (not
$2,500.00)
The airfare was *$250.00* and the lunch cost *$2.25.*

Numbers should not be next to each other. If this becomes
necessary, separate them with a comma.

BAD EXAMPLE: In 1975, 5 studies were performed.
REVISION: Five studies were performed in 1975.
REVISION: In 1975, five studies were performed.

Do not separate units of measurement between one line and
the next.

EXAMPLE: The pump produced only 45
gpm.
REVISION: The pump produced only 45 gpm.

Use commas to separate thousands when numbers run to
four or more figures. Line up the commas if the numbers
are in a column.

EXAMPLE: 7,325
             420,029
           1,862,777

Note that in Europe and in much of the rest of the world,
the commas and periods are reversed when used in
numbers. In Europe, for example, $3,78 would mean three
dollars and seventy eight cents.

Do not imply greater accuracy than the data provide. Note
that zero (0) is different from 0.00. The latter implies a
level of sensitivity that may not have been intended. If a
higher sensitivity is available, it may be that 0.00 is actually
0.004, whereas zero (0) means *nothing*. If you say that the

dissolved oxygen level in the aeration tank was *0.00* mg/L you mean that it actually was less than 0.004 mg/L and that the number was rounded off to 0.00 mg/L. This is different from *0 mg/L*, which implies that there was *no* oxygen in the aeration tank.

EXAMPLES: *2.3 mg/L* is not the same as *2.30 mg/L*
93 percent is not the same as *93.0 percent*

Capitalize a noun followed by a number or a letter that indicates sequence. However, do not capitalize *line, note, page, paragraph, size,* and *verse.*

EXAMPLES: Table 4; Figure A; Section 4; page 35; paragraph 3; Pump 3; Grid 2A

The word *number* or the abbreviation *No.* is usually unnecessary when referring to a numbered item.

EXAMPLE: Refer to Invoice 405, Item 35, page 2

In certain contexts, however, it may be necessary to use the word *number* or the abbreviation *No.*

EXAMPLES: Valve *No. 3* ... The correct *number is 37..*

When typing dimensions, the symbol x is not a substitute for the word *by.*

EXAMPLES: Use 8 1/2- *by* 11-inch paper.
The paper measures 8 1/2 *by* 11 inches.

2.1.6 Abbreviations

Avoid abbreviations in engineering writing. If abbreviations are used, they must be consistent throughout the document. For example, if the United States Department of Commerce is abbreviated as U.S. DOC in one part of the text, it is inappropriate to abbreviate it as D.O.C. or DOC later on.

Here are some guidelines on the use of abbreviations:

Do not abbreviate terms and units of measurement in non-technical documents. Spell out all terms the first time they occur in a letter or memo or the first time they appear in each chapter of other documents. Place the abbreviation in parentheses following the term.

EXAMPLE: ... 23 *million gallons per day (mgd)*.

Following such an explanation, use of the abbreviation is permitted. If, however, there is any chance whatsoever that the reader may not remember your shorthand, repeat the definition.

Avoid uncommon abbreviations such as *gallons per capita per day,* usually written in engineering reports as *gpcd*. If you saw this abbreviation for the first time and had no idea what it meant, you could never figure it out. Make sure you do not ask your readers to be mind-readers.

The ampersand (&) is not an abbreviation for the word *and* but rather a symbol sometimes used in lieu of the word *and*. Use an ampersand when typing firm names only if an ampersand appears in the official name of the firm.

Do not abbreviate words such as *street, avenue,* and *boulevard.* Spell out all place names, such as Peachtree

Street and Nile River. Two-letter abbreviations of state names is acceptable.

Do not use periods after letters in abbreviations of association names, names of certain government offices and agencies, or names of other official bodies. An exception is U.S. for the United States.

> EXAMPLES: *ASCE* (not *A.S.C.E.*), *U.S. EPA* or *EPA* (not *U.S.E.P.A.* or *E.P.A.*)

Some states also have environmental protection agencies, and if both the state and federal agency are referred to, the letters *U.S.* are essential in identifying the federal EPA.

When abbreviating United States as part of an organizational name, use a space only after the final internal period.

> EXAMPLE: *U.S. Postal Service*

Do not substitute symbols for words such as *percent, degrees, pounds, number, at, inches, foot,* or *feet*.

Do not abbreviate the words *figure(s)* and *table(s)*.

Make a capitalized abbreviation plural by adding a lowercase *s*.

> EXAMPLE: *BODs, SOQs*

The addition of *s* does not normally make an abbreviation plural. The plural of *lb* is *lb* (not *lbs*); the plural of *kv* is *kv* (not *kvs* or *kv's*).

Do not abbreviate units of measurement of six letters or

less (for example, *inches, feet*).

Abbreviate units of measurement only if preceded by numerals.

EXAMPLE: . . . 3400 cfs . . .

Do not use periods after scientific or technical abbreviations.

2.1.7 References and Footnotes

References and footnotes are used in engineering reports to give additional evidence and support for a statement, or to give the source of a fact or quotation. A bibliographic reference must supply sufficient information to direct the reader to the original source document.

Three styles are acceptable for references. The first style numbers all references consecutively as they appear in the text. Thus the first reference encountered is number "1", usually as a superscript. The references in the list of references are then numbered consecutively.

EXAMPLE: As noted Riley and O'Brien,[1] the pump ...

The second technique lists all the references in the list of references in alphabetical order based on the author's last name. The reference numbers in the text then correspond to the reference list.

A third method uses the author's last name and publication date in parentheses. For example, the reference might appear in the reference list as:

Jones, A. K., (1990) "Thickening Characteristics of Primary Sludge," *Journal of the Environmental Engineer Division*, ASCE, v. 112, n. EE4, pp. 87-94.

In the text this article would be listed as (Jones 1990). The primary advantage of this technique is that revisions do not require reference renumbering. The reference list is arranged alphabetically by author's last name, unnumbered, and revision involves simple additions or deletions.

The last style is now favored by numerous technical journals. ASCE, for example, uses this style but insists that references be in the following order: last names and initials of the authors, the year of publication (in parentheses), the title of the paper (in quotes) or title of book (in italics), volume number, issue number (in parentheses), publisher, city and state or nation of publication, inclusive page numbers, and a period.

There are many styles for writing references, and some editors will be extremely sensitive to breaches in a required style. But the truth is that a comma dropped here or there will not make any difference as long as this is consistently done and there is no ambiguity.

Below are the requirements for articles and books published by the American Society of Civil Engineers (ASCE). Details may be found at www.asce.org. Most other engineering societies have similar requirements for listing references.

Journals: King, S., and Delatte, N. J. (2004). "Collapse of 2000 Commonwealth Avenue: Punching shear case study." *J. Perf. Constr. Facil.,* 18(1), 54-61.

Note that the volume of the journal comes first (18) followed by the number (1) and then the pages (54-61).

Conference Proceedings and Symposia: Fwa, T. F., Liu, S. B., and Teng, K. J. (2004). "Airport pavement condition rating and maintenance-needs assessment using fuzzy logic." *Proc., Airport Pavements: Challenges and New Technologies*, ASCE, Reston, Va., 29-38.

74

Books and Reports: Feld, J., and Carper, K. (1997). *Construction failure,* 2nd Ed., Wiley, New York.

Unpublished Material: Unpublished material is not included in the references but may be cited in the text in the following forms: (John Smith, personal communication, May 16, 1999) or (Jones et al., unpublished manuscript, 2002).

Web Pages and On-line Material: Burka, L. P. (2002). "A hypertext history of multiuser dimensions." *MUD history,* <http://www.ccs.neu.edu> (Dec. 5, 2003).

The date in the end is the last time the site was accessed.

CD-ROM: Liggett, J. A., and Caughey, D. A. (1998). "Fluid statics." *Fluid mechanics* (CD-Rom), ASCE, Reston, Va.

Reference lists are prepared in the same format whether they appear as footnotes at the bottom of pages, at the end of a chapter, or at the end of a report. The placement of references within a report depends on the total number of references cited regardless of the report's size.

If there are many references in a report they are commonly placed at the conclusion of each chapter. If there are only a few references, they can be listed in the appendix or presented as footnotes. If an appendix is used to list all of the references, a separate list must be used for each chapter. If only one or two references are cited in a letter report, footnotes are commonly used.

2.2 WRITING INSTRUCTIONS

Some of the worst engineering writing is the writing of instructions. Engineers are notoriously unable to explain things in words, mainly because whatever is to be explained is so obvious to them that they cannot understand why someone would have difficulty understanding it. (An alternative explanation is that too many engineers are illiterate!) In writing instructions, if the reader misunderstands the instructions, *it is the fault of the writer,* and thus the onus is on the writer to be so clear that mistakes by the reader are impossible.

Here are several helpful hints about writing instructions:

Use the active voice. Especially when writing instructions, the words have to be simple and concise. Always use the active voice because it is the simplest form of speech.

> BAD EXAMPLE: The board should be cut into two pieces.
> EVEN WORSE BAD EXAMPLE: It is necessary to cut the board into two pieces.
> REVISION: Cut the board into two pieces.

The first word, "cut," is the operative word.

Write instructions in chronological order. Do not place a note later in the instructions referring to what should have been done earlier. Include adequate precautions and warnings within the instructions. Do not depend on blanket warnings covering the entire procedure.

Define terms. Something that might be well known to you might be a total mystery to the reader. Even such simple things as the difference between bolt, screw, nail, carriage bolt, molly bolt, etc. need to be made explicit, usually with pictures.

2.3 ADAPTING YOUR WRITING TO THE AUDIENCE

Adapt your writing to the audience and purpose. In most

writing, don't use long words and sentences where short ones will suffice. However, formal reports often demand the use of more formal language. For example, in an informal note, you might refer to some rooms as "labs," but if writing a technical paper you would call these rooms "laboratories." Similarly, avoid a folksy tone in documents that demand formality.

One way to gauge the level of your writing is to measure the length of words and sentences and calculate the *Fog Index*, a quantitative measure of the readability of writing.[3] A very high Fog Index means that the writing may be impenetrable, while a very low Fog Index implies that the writing is simplistic. You can calculate your own Fog Index in this way:

1. Choose an excerpt of at least 100 words from something you have written.
2. Determine the average number of words per sentence (count clauses separated by colons and semicolons as sentences).
3. Count the total number of words in the passage.
4. Calculate the percentage of big words, three syllables or more. Don't count proper nouns, compound words made up of simple words such as "sunstroke," and all verbs made into three syllables by adding "-es" or "-ed."
5. Average the two figures (Add the number of words per sentence and the percent of big words, and divide by 2). The resulting number is the Fog Index.

The example below shows how the Fog Index is calculated.

EXAMPLE:
Calculate the Fog Index for the following paragraph written by Ralph Waldo Emerson:

The greatest delight which the fields and woods minister is the suggestion of an occult relation between man and the vegetable. I am not alone and unacknowledged. They nod to me, and I to them. The waving of the boughs in the storm is new to me and old. It takes me by surprise and yet is not unknown. Its effect is like that of a higher thought or a better emotion coming over me when I deemed I was thinking justly or doing right.[4]

The average number of words per sentence:
(21+6+8+14+10+26)/6 = 14.2.
Total number of words: 85
Total number of big words: 5
Percent of big words: (5/85) x 100 = 5.9.
Average (5.9 + 14.2)/2 = The Fog Index = 10.

What is the proper Fog Index? It depends on the audience. *The Wall Street Journal*, *Time*, and *Newsweek* all have a Fog Index of about 11. A typical technical journal may have a Fog Index of about 20, and a sixth-grade reader is around 6. Ralph Waldo Emerson (from the limited quotation in the above example) has a Fog Index of 10.

Some word processing programs now provide a grammar check that often includes a world count and a calculation of complexity of the writing. This is a useful tool for checking your level of presentation.

Finally, make sure you use an appropriate tone in correspondence. Negative-sounding letters can be especially counterproductive. Caskey[5] points out that even such benign words as *obviously* or *claim* can take on a negative tone.

BAD EXAMPLES: Obviously you know that the bill is overdue.
You failed to give us the flow data.

These sentences imply stupidity, ineptness, or lack of care.

78

BAD EXAMPLE: You claim you sent us the data.

This sentence implies that the other party lied, and even if it is not meant to be insulting, it is another example of unfriendly tone.

2.4 THE USE OF HUMOR

Generally speaking, humor is neither expected nor appreciated in engineering writing. The only place humor might be appropriate is in informal correspondence within a firm, or with clients/customers who are also very good friends. Even then some types of humor, such as political or religious humor, can be problematical. It is best to avoid it altogether in engineering correspondence. *

REFERENCES

1. Franklin, B., *Pennsylvania Gazette*, August 2, 1733, as quoted in Andrews, D.C., and Blickle, M. D., *Technical Writing; Principles and Forms,* Macmillan Publishing Co., New York, 1982.
2 Strunk, W. Jr., and White, E. B., *The Elements of Style,*

---

* One of the problems with humor is that it is not easy to indicate that something in a written document is supposed to be funny. In speaking, one can use body language such as a smirk or twinkle to assure the listener that the remark is meant in jest, but the author of a written work has no such opportunity. How is the reader to know that something is to be taken humorously?

Of course, one way to accomplish this is to use a "humor mark", much the same as we use a question mark or quotation mark. Unfortunately, no such mark is widely recognized. E-mail correspondence has evolved its own humor marks such as *lol,* :) , or ☺ , but these should not be used in engineering correspondence.

Second Edition, Macmillan Publishing Co., New York, 1972.

3. Gunning, R., *How to Take the Fog Out of Writing*, The Dartell Corporation, Chicago, IL, 1964, as quoted in Ulrich, G. D., "Write a Good Technical Report," *Chemical Engineering*, McGraw-Hill Publications, September 5, 1983.

4. Emerson, R. W., "Nature" in *The Best of Ralph Waldo Emerson*, Walter J. Black, New York, 1941.

5. Caskey, C.O., *Frugal Me! Frugal Me! A Light Approach to Serious Writing*, R. L. Bryan Co., Columbia, SC, 1985.

PROBLEMS

2-1. Rewrite the following:

a) The unlikelihood of meeting orders from the majority of its customers is of concern to the company.

b) The possibility of the situation facilitates the relationship to the extent where concepts will purposely substantiate the system in a manner congruent with the conceptional standpoint.

c) A test was made to verify the hypothesis.

d) Procedures were initiated to reduce the accidents.

e) The aeration basin utilizes a floating aerator.

f) Following the sacrifice of the test species, the tests were terminated.

g) The size of the magnitude of the pile of wood chips was awesome.

h) The optimum utilization of time was conceptualized.

i) The students fabricated the pilot plant.

j) This was conceptualized as a viable alternative.

k) The tests showed uncertain success.

l) All available evidence seems to point toward the inescapable conclusion that the mice liked the oatmeal.

m) He substantiated the suggestion that someone should make a study of space utilization.

80

n) Instructorwise, this is a dull class.
o) If the beam failure shows the desired effect, it will be automatically recorded.
p) This study, which is as yet inconclusive, supports the theory, within prescribed limits, that the apparent, though yet unmeasureable, coagulation of the colloidal particles, though in no other observable test result, is due, to the best of our knowledge, to the increased salinity, or what may be termed the increased dissolved solids in the water.

2-2. Edit the following:

This homework assignment utilizes a viable methodology, technologywise, being written on the computer in the laboratory and hopefully the testee's attempts to arrive at an approximation of the magnitude of the correct responses are not subject to the differences in the variabilities of the typing techniques or software, possibly.

2-3 Using all of your descriptive and observational skills, *write* directions on how to travel from your room to the classroom. Assume that this will be used by a stranger to your campus. Then draw a map for the same route. Compare the two. Which would be easier for the stranger? Why?

2-4 Simplify the following phrases:
1. at the present time
2. during the time that
3. to make a study of
4. to take into consideration
5. in the neighborhood of
6. due to the fact that
7. on account of the fact that
8. In the event that

9. owing to the fact that
10. in the order of about
11. a large number of
12. be of the opinion that
13. during the course of
14. in a manner similar to
15. An attempt to arrive at an approximation as to the magnitude of
16. eradicate completely
17. free gift
18. has the capability to
19. give assistance to
20. are applicable
21. kills bugs dead in six weeks
22. center around [think geometry here]
23 completely eliminate
24. the frustrated squirrel hunter [two possibilities]
25. the old Indian fighter [two possibilities]
26. true facts
27. basic fundamentals
28. on a daily basis
29. by means of
30. in the area of
31. make contact with [careful!]
32. make a purchase
33. give approval to
34. have a deleterious effect on
35. have a tendency to
36. have an influencing effect on
36. tend to impact
37. until such time that
38. in the near future
39. past history
40. . . . final end.

82

2-5. Tear out one page of your local newspaper's real estate section and read each notice. Underline in red those words that are euphemistic, misleading, or just plain silly. List these and number the times they are used.

2-6. *So You Think You're a Pretty Good Speller Department.*
Check the following sentence for spelling: "Neither financier niether siezed nor recieved either speceis of weird liesure wiers."

2-7 *Doublespeak Department.*
Listed below are some famous and not-so-famous phrases which we now recognize as doublespeak, language that pretends to communicate but is in fact the worst kind of obfuscation – language that misleads, shifts responsibility, or makes bad seem good. Study the doublespeak phrases, jot down your ideas of what they really mean, and be prepared to discuss them in class.

    a)   for your convenience
    b)   postage and handling
    c)   pre-owned cars
    d)   negative patient care outcome
    e)   periods of negative accelerated economic growth
    f)   poorly buffered precipitation
    g)   wood interdental stimulator

2-8 Edit the following:
a) And the itching with a desire to scratch and rub the area affected persisted throughout the second day.
b) Although the polymer is of the cationic type, it did not increase the size of the flocs.
c) We must run this calculation in order to determine whether or not this reaction could be utilized for practical design purposes.

d) The majority of the cracks encountered were of the shear failure kind.

e) The engine required 2 7 inch cams.

f) In general, it would be expected that students would do little work during the last days of school.

g) On the tenth day the rats were sacrificed.

h) The report was due on the 25th of June last year.

i) The data is shown in Figure 6.

j) Table 5-8 shows how temperature impacted the speed.

k) Engineers are well schooled in basic fundamentals.

l) The solar cell environment energy conversation efficiency improvement considerably increased when quartz shields are placed over the solar cells.

m) This firm makes cans for shortening people.

n) The population is centered around the business district.

o) It is obvious to the most casual observer that the differential equation can be solved to yield . . .

p) It is believed that the Civil Engineering Department should . . .

q) Three experiments were conducted, and the data substantiated the theory except on the second and third tests.

r) Lets technicalize the report.

s) It is essential that the course be modified.

t) Please bring the instant coffee heater to the meeting.

u) The nature of helium is such that it is a gas at room temperature.

v) The recent softness in the home-building field has hit the plywood industry hard.

w) The population of California in 1960 was approximately 8,956,834.

2-9 The following extract comes from an actual report by a dean. (Honest!) Translate it into English.

What has all this got to do with graduate education? To some this discussion of the labor force may seem irrelevant with respect to the more immediate and specific problem of securing appropriate jobs by new Ph.D.'s in mathematics, physics, history, sociology, etc. My point is that it is indeed relevant. The most highly educated component of the labor force is after all only a component. Utilization of that component is dependent in the most fundamental way on the deployment of the entire labor force. I reiterate my remarks about the overriding importance of the patterns of national priorities and the overall state of the economy. With respect to the future of graduate education one can extract reassurance from this description of the labor force. A sophisticated and highly differentiated occupational structure is the natural habitat of the highly trained person. As the educational and skill level of the general labor force continues to be upgraded, concomitantly with the changing requirements of that structure, a more favorable environment is continually being constructed for those who have received the highest levels of training in the fundamental areas of knowledge. This is in long-run defiance of the short-run exigencies of the marketplace and temporary imbalances between supply and demand for particular types of personnel.

2-10 Why are the following opening statements inappropriate for memos? Suggest alternatives.

a) You obviously meant to send me the data you promised, but simply forgot.

b) In your memo of 12 November, you made some erroneous statements that anyone familiar with the project would not have made.

c) I hoped that we could complete this job by last week,

but you failed to contact Tom Brown.
d) You claim that the best route for us to follow is to sell the Duquesne Light stock.

2-11 Edit the following abstract and weak statements into concrete, specific sentences. Use your imagination.
a) The weather was nice.
b) The test took a long time to run.
c) The chances of this landfill liner failing are slim.
d) The building was tall.
e) Since he came from another university, he could not have known about the local beer drinking games.
f) I dislike the teacher so much that I believe I will have to talk to my advisor about doing something.
g) My roommate is reading a book.

2-12 Why are these phrases redundant? Explain and revise.
a) united coalition
b) tuna fish
c) totally independent
d) true facts
e) software program
f) 6 am in the morning
g) seen by the eyes of
h) reiterate again
i) random chance
j) RAM memory
k) please RSVP
l) foreign imports
m) final conclusion
n) empirical observation
o) dead corpse
p) cooperate together
q) consensus of opinion
r) foot pedals
s) equal halves

86

t) DOS operating system
u) central core
v) AC current
w) LCD display
x) NIT tournament
y) Open Seven Days a Week and on Weekends
z) No Trespassing Without Permission

2-13 Improve the following sentences. Submit your best alternative.
a) Ask not for whom the bell tolls. It tolls for thee.
b) I have but one life to give to my country.
c) I think, therefore I am.
d) Damn the torpedoes. Full steam ahead.
e) This day will live in infamy.

2-14 Are any of the following a misuse of numbers? If so, how and why?
a) The average class size is 34.5 students.
b) The elevation of Mt. Washington is 6348 feet.
c) He must have been going at least 72 miles per hour when he passed us. [Careful on this one!]
d) There are 11 players on the soccer team.
e) The average family has 2.7 children.
f) The ant colony had 64,867 ants.
g) The weight fraction of glass in the garbage from this town is 12.358%.
h) The "Big Dig" in Boston cost the taxpayers $126,453,234.67.
i) A new tank for the U. S. Army costs $1.26 million.
j) We estimate that the installation of air conditioning for the school in Frog Pond, NC, would cost about $3,800.
k) The lowest bid for the installation of air conditioning in the Frog Pond school came in at $3,446.
l) The grass in my front yard grows at a rate of 0.000034468 km/s.

CHAPTER 3

THE USE OF TABLES AND GRAPHS IN ENGINEERING WRITING

Effective engineering writing requires informative, complete, and memorable displays of quantitative information. Quantitative information is most often expressed in either tables or in graphs.

3.1 TABULAR DISPLAY OF QUANTITATIVE INFORMATION

Tables should always be self explanatory, with a complete title explaining what the table is all about. Because tables are often copied from articles and distributed without the accompanying text, they should be able to stand alone, without benefit of the text. The opposite is not true, however. The text needs the table, and all tables (and figures) have to be "called out" in the text.

The placement of tables in the text is also important. Whenever possible, tables should be constructed so that the reader does not have to turn the page in order to read the table.

Suppose it is necessary to tabulate some temperature data at different depths in a lake. Four different tabular arrangements are illustrated in Tables 3-1 through 3-4.

Table 3-1 is nothing more than tabular data in prose form, very difficult to interpret, and without the usual advantages of tables. The title of Table 3-1 is quite uninformative. Everyone can see that the table has depth vs. temperature. But what temperature, where, and when?

88

Table 3-1
Temperature vs. Depth

D(depth)=0, T(temperature)=20.12°C; D=1m, T=19.34°C;
D=2m, T=18.56°C; D=3m, T=17.78°C

Table 3-2 is incorrect since it places the *dependent* variable first. This is not how people think when reading scientific literature. Tables should be arranged with the *independent* variable in the far left column. The title is much improved, because now we know that these are water temperatures in Lake Michie, but we still do not know where or when these were taken.

Table 3-2
Temperature Soundings for Lake Michie

Temperature (°C)	Depth (meters)
20.12	0
19.34	1
18.56	2
17.78	3

The title of Table 3-3 is an improvement, and this table as it stands could be removed from the text and it would still have validity. The data in Table 3-3 are arranged so that the independent variable is to the left and the dependent variable is to the right. But in this table the depth decreases as one scans the table downward, resulting in a double-take on the part of the reader, and again encourages misinterpretation.

Table 3-3
Temperature Soundings for Lake Michie
(Station 6A, June 13, 1939)

Depth (m)	Temperature (°C)
3	17.78
2	18.56
1	19.34
0	20.12

Table 3-4 (finally) is arranged in a way that is most comfortable for the reader and has the least possibility of misunderstanding.

Table 3-4
Temperature Soundings for Lake Michie
(Station 6A, June 13, 1939)

Depth (m)	Temperature (°C)
0	20.12
1	19.34
2	18.56
3	17.78

In the text, Table 3-4 might be referred to in the following way:

The temperature of Lake Michie has been steadily increasing. For example, Table 3-4 shows the temperature soundings for Station 6A (close to the bridge) taken in 1939. By contrast, the temperature soundings taken in June of 2009 are shown in Table 3-5.

Tables do not show relationships but they do present data in a form that allows the reader to make inferences. For example, in Table 3-4 the numbers show that the temperature decreases with depth.

This information could have been presented equally well in graphical form, but information would have been lost because it would not have been possible to show the temperatures to four significant figures.

3.2 GRAPHICAL DISPLAY OF QUANTITATIVE INFORMATION

In engineering and science, as well as other disciplines such as history, visual forms can often convey quantitative information better than can verbal forms. In this section we discuss various forms of illustrations used in such displays.

3.2.1 Line Graphs

The most popular illustration of quantitative material is the line graph. A line graph shows how one variable relates to another variable by plotting the data along a vertical axis (the *ordinate*) and horizontal axis (the *abscissa*). The purpose of a line graph is to show the relationship between these two variables. Generally, the dependent variable is plotted on the vertical axis and the independent variable lies on the horizontal axis. For example, suppose you conduct an experiment to see if study time has any effect on test grades. You plot the time spent studying for tests on the abscissa (horizontal axis) and the resulting test grade on the ordinate (vertical axis). The resulting data points might look like Figure 3-1. What information can we obtain from this graph? We might conclude that the data show a distinctly positive effect of studying on grades, with more study time resulting in better grades, and we might be tempted to draw a line through the data indicating this trend.

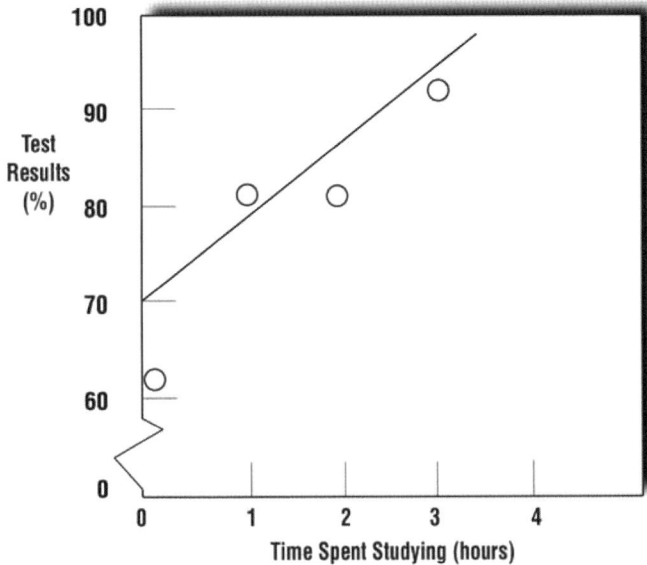

Figure 3-1 Correlation between study time and test scores.

But what if the data had turned out like Figure 3-2? Does this indicate the opposite – that increased study time results in lower test scores? If this is not the explanation, what might be the reason for the curious data? It could be, of course, that you intentionally manipulated the data, or it could have been that the courses which you find difficult are the ones you study most for, and often get the worst grades in. It does not make sense therefore to presume a causal relationship here just because there is a correlation. Although line graphs are useful pictures of quantitative information, we have to be careful how they are interpreted.

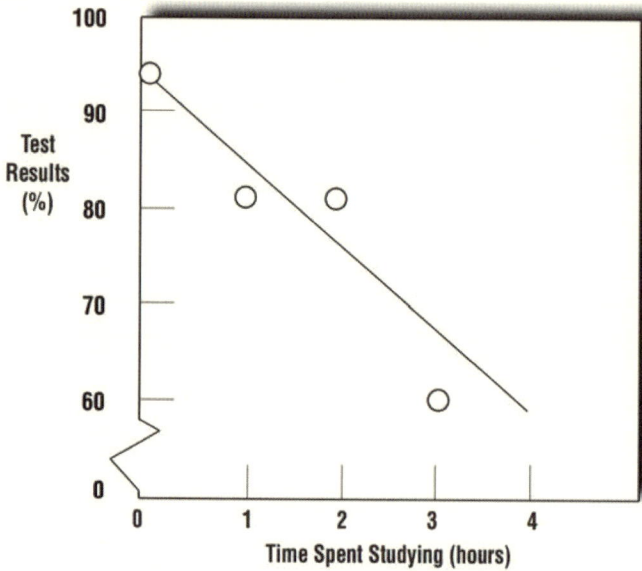

Figure 3-2 Alternative correlation between study time and test scores.

The following are some pointers on drawing good line graphs:

Title each line graph. The titles should provide a complete explanation of what the data are and where they originated. Every graph must be self explanatory and stand on its own if separated from the text.

Cite all graphs in the text. If there is no need to reference the graph in the text, there is no need for the graph.

Limit the amount of material on each graph. A line graph that is too crowded takes too long to decipher, and the reader might give up. Make the data points heavy, and clearly distinguish among different data sets. Remember that the objective of graphical display of quantitative information is speed and clarity.

Orient the graph appropriately. The graph should fit on the page right side up so the reader does not have to turn the book or report sideways. If this is not possible, the graph must be read from the right (outside of the report if it's on the right hand page).

Organize the variables correctly. The independent variable is the abscissa (horizontal axis) and the dependent variable is the ordinate (vertical axis). The only exception to this rule is when the independent variable is elevation (depth or height) since it makes little sense to place depth or height on a horizontal axis.

Begin the scales at zero. If it is inconvenient to begin a graph at zero, a broken scale must be used. A broken scale is indicated by a squiggly line or two parallel lines through the axis. Figures 3-1 and 3-2 illustrate how this is done.

Use simple multiples on the scales. Divide the scales into simple units such as 2s, 5s, 10s, or similar easily added and interpolated numbers.

Connect the data points if this is reasonable. Often it is better to show a trend and not connect the points. If the data are inadequate, or if extrapolation is needed, use a dotted line to show that sufficient data are not available. A solid line indicates that you are confident in the relationship but are unwilling to conclude that the line shows the accurate values. When data are used to extrapolate into the future (such as population growth), the extrapolated line must be a dotted line since it is not based on actual data.

Label the curves if the graph has more than one line. If this is too cumbersome, use a key. Figures 3-3 and 3-4

94

show how labels and a key can be used to express the same data.

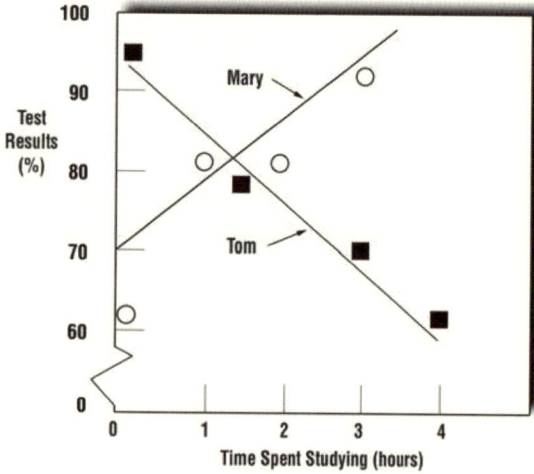

Figure 3-3 Identification of two different data sets on the same plot by labeling the curves.

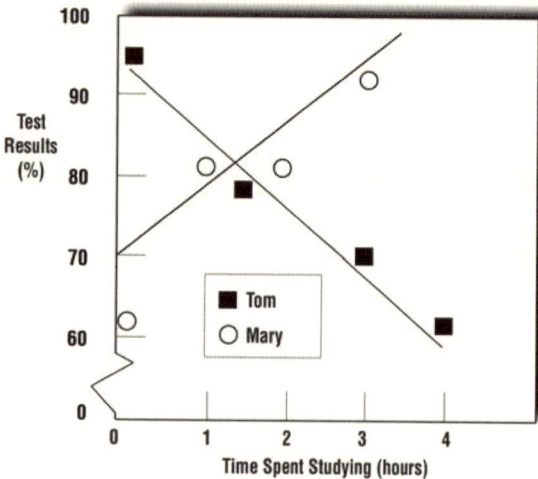

Figure 3-4 Identification of two different data sets on the same plot using a key.

When a line graph illustrates multiple dependent data points for a few independent variables, one means to accurately and effectively present the data is to graphically plot statistical information instead of the individual data points. Figure 3-5 shows one such method, where the mean is shown as a line in the middle of the box. The box itself represents the distribution of data – plus and minus one standard deviation from the mean. The vertical lines extending from the top and bottom of the box show the 95% confidence interval for the data. Such a plot shows much more information than a graph that simply plots all of the data points.

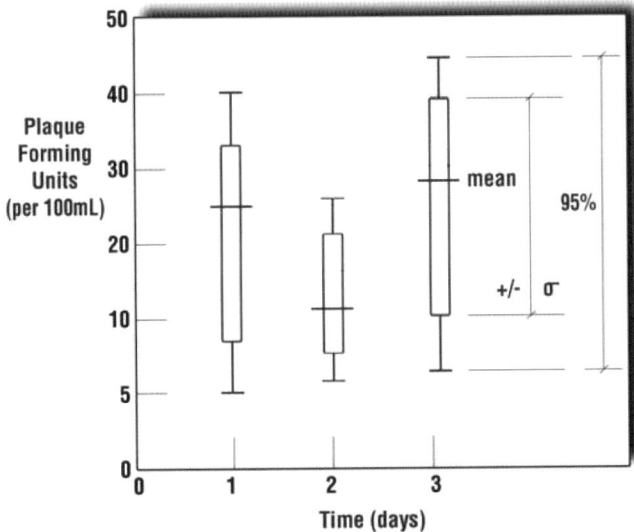

Figure 3-5 A method for illustrating multiple dependent data points for limited independent variables. The box represents the range of data within one standard deviation on either side of the mean, illustrated by the line through the box. The tails show the 95% confidence intervals for the data.

96

3.2.2 Histograms

Histograms, or *bar graphs*, are useful for illustrating numerical lists, especially when the independent variable is not a continuum. Suppose you want to illustrate the number of various colored balls found in a barrel. Because color is not a continuous variable, a histogram is appropriate. Figure 3-6 shows both a correct bar graph and an incorrect line graph. The line graph is inappropriate because letter grades for a course do not represent a continuum; grades are discrete and therefore require a histogram.

3.2.3 Other Illustrations

In addition to the bar graph and the line graph, engineers and scientists can make use of numerous other illustrations. For example, Figure 3-7 shows a *spot map*, a common city street map with hash marks showing the location of some event of significance. Figure 3-7 is, in fact, a section of the very first spot map ever drawn and represents a watershed event in public health.

During the middle of the nineteenth century, the transmission of infectious diseases was still a mystery. Pasteur and Koch had not yet identified microorganisms as the culprits. Various theories abounded, including the miasma theory, which blamed night air for the transmission of such diseases as cholera, typhoid, and dysentery.

In 1852, a particularly vicious cholera epidemic struck London. John Snow (1813-1858), a public health physician, observed that the incidence of cholera in London followed a curious pattern. He noted that the locations of the homes of people stricken with cholera seemed to center on the public water supply pump on Broad Street. He drew the

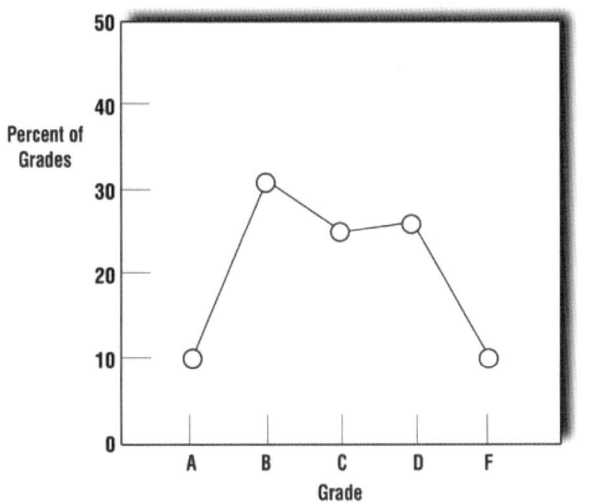

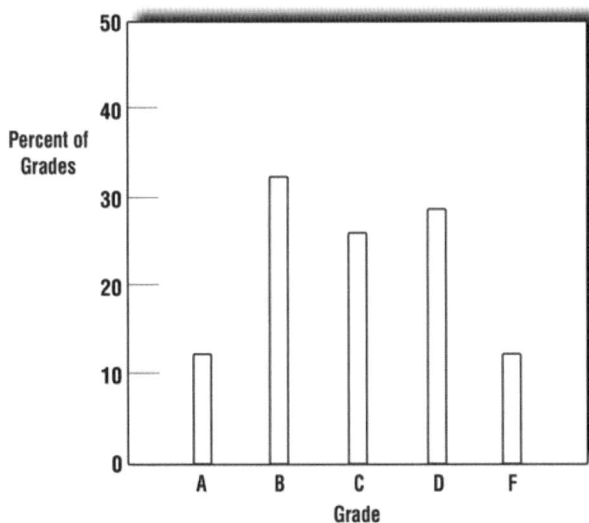

Figure 3-6 A correct histogram, and an inappropriate use of a line graph.

98

spot map in Figure 3-7 to show the locations of all incidence of cholera, and based on such graphical demonstration, convinced the City of London to remove the pump handle from the pump. This prevented the use of the contaminated water and the epidemic subsided.

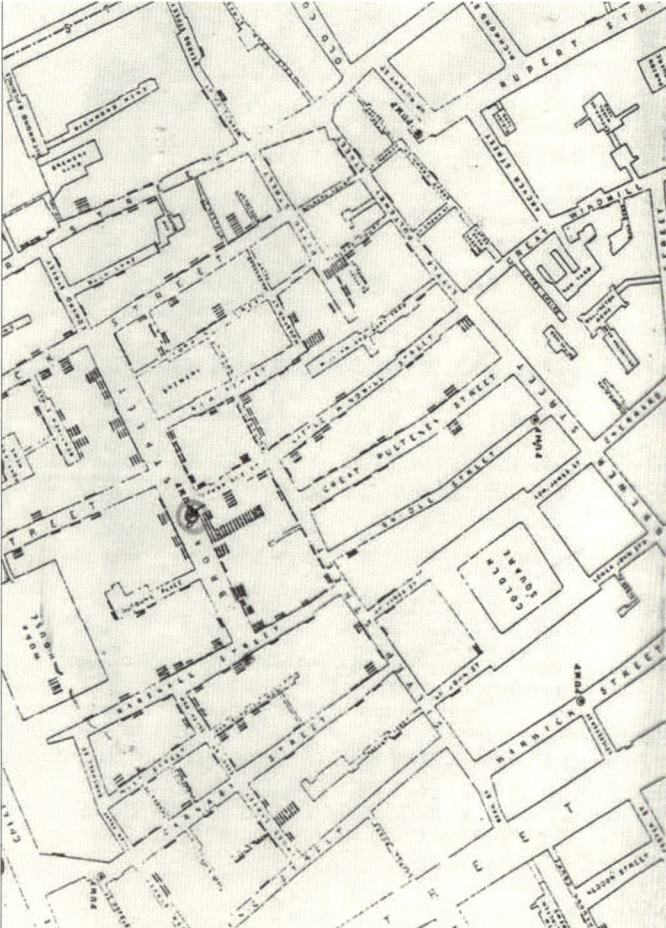

Figure 3-7 Part of John Snow's spot map showing the incidence of cholera during the 1852 epidemic in London. The water pump on Broad Street is indicated on the figure by a bull's eye.

Another way to illustrate data when time is a variable is the *time series graph*. One of the most famous time series graphs was drawn by Charles Minard (1781-1870). The graph depicts the plight of Napoleon during his invasion of Russia. Figure 3-8 tells the whole gruesome story of Napoleon's debacle and is worth a close study. The striped bar shows the size of Napoleon's army as it traveled to Moscow, and the solid bar shows his army as it returned to France. This graph is a history lesson all by itself.

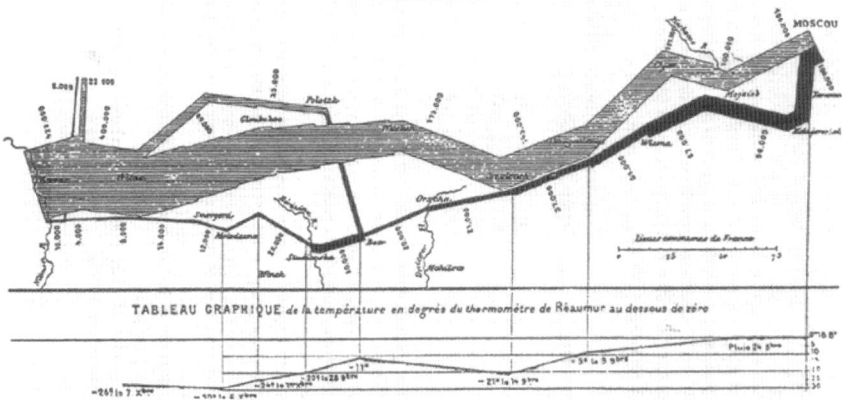

Figure 3-8 Graphical representation of Napoleon's Russian campaign, 1812-1813. (Marey, E. J. *La Methode Graphique, Paris 1885,* reproduced in Tufte. E. R. *The Visual Display of Quantitative Information* Graphics Press, Cheshire, CN, 1983. Used with permission)

The point of this discussion is that all illustrations do not have to be simple bar graphs or line graphs. Imagination and originality are positive attributes in illustrations just as they are in writing.

PROBLEMS

3-1 Collect typical graphics from any issue of *USA Today*. Reproduce them and discuss their effectiveness and integrity.

3-2 Construct a line graph for the following data. Be sure to follow all the principles of graphical presentation and include a proper title. The data are from an experiment in which three types of microorganisms were grown, Strain A, B, and C, and their growth was recorded as the number of microorganisms per 100 mL. The reactor was inoculated (time zero) with 10 microorganisms per 100 mL.

	Concentration, number/100 mL		
Time, days			
	Strain A	Strain B	Strain C
1	20	33	42
2	40	63	55
3	78	126	63
4	152	142	68

3-3 Which of the following would you plot as a line graph and which as a histogram? Draw the graphic you would choose and be prepared to argue for your choice. You don't need real numbers; just make something up.

a) The number of students receiving various grades (A,B,C,D, and F) in a class.
b) Your grade point average over all the semesters in college.
c) The average grade point average at the University over the last ten years.
d) The average cost of real estate in your community over the past ten years.
e) The number of B-2 Stealth bombers manufactured by

Boeing every month during 2005.
f) The number of 1956 Ford Thunderbirds registered in the United States since 1956.
g) The cost of first class postage over the past 100 years.
h) The available memory provided on typical lap top computers over the past five years.
i) Number of Nobel Peace prizes awarded to citizens of various countries.
j) Penetration depth of a pile as a function of the number of blows by the hammer.

3-4 Construct appropriate graphical representations of the following data sets. Some of the data sets might require more than one graphic. Use your imagination.

A. The time spent by a Finnish farmer during the year.

Season	Hours in a day spent on			
	Farm work	Eating	Sleeping	Recreation
Spring	12	2	8	2
Summer	14	2	4	4
Fall	12	2	6	4
Winter	4	3	10	7

B. Farm exports from the three Baltic countries in 2008.

Product	Estonia	Latvia	Lithuania
		Tons per year	
Butter	405	126	289
Cheese	125	300	118
Barley	308	80	40
Oats	0	140	140

C. The arrival times of trains at Victoria Station during June 17, 1965, from midnight to 8 a.m.

Train	Scheduled Arrival	Actual Arrival
A	0:12	0:12
B	0:45	0:42
C	1:15	1:12
D	2:05	1:55
E	5:45	5:48
F	6:05	6:10
G	6:18	6:25
H	6:36	6:35
I	6:47	7:12
J	7:18	7:22
K	7:35	7:45

3-5 Sketch an appropriate graphic for the following data obtained for a trolley line on Liberty Avenue in Pittsburgh during the run of Car 34 on September 26, 1970:

Station	Number of passengers	
	Embarking	Disembarking
0 (rail yard)	6	0
1	26	0
2	47	12
3	89	49
4	46	57
5	12	6
6	6	12
7 (end of line)	0	96

3-6 Sketch an appropriate graphic for the following data obtained in measuring the head loss and velocity of oil when pumped in a 200 foot long, 6 inch diameter PVC pipe:

Velocity (ft/sec)	Head Loss (ft)
0	0
1	0.7
2	2.4
3	4.7
4	7.7
5	11.5
6	16.0

3-7 Sketch appropriate graphics for the following data:

Color of cars	Number of cars manufactured in 1980	Number of cars still on the road in 2000
Black	10,400,000	2,600,000
White	6,600,000	2,100,000
Silver	5,300,000	1,200,000
Blue	12,300,000	1,800,000
Red	3,300,000	260,000

What conclusions can you draw from your graph? Is there cause and effect?

3-8 If you have to explain to someone where on your campus you have had classes (which buildings), how would you do this graphically? Submit an appropriate graphic.

3-9 As an experiment at Podunk College, a single term paper was given to 20 professors for grading. Two of the characteristics of the professors and the grades they gave are shown below:

Professor	Age	Discipline	Grade
1	54	Engineering	C
2	67	History	C
3	33	English	D
4	33	Engineering	A
5	45	History	C
6	66	Psychology	B
7	51	Engineering	C
8	43	Sociology	D
9	44	Engineering	F
10	38	Music	C
11	39	Engineering	A
12	67	Economics	B
13	83	Art History	F
14	27	Classics	B
15	36	Romance languages	A
16	58	Physical education	C
17	45	Chemistry	B
18	49	Physics	C
19	62	Biology	B
20	74	Geology	F

How can you best represent these data graphically? It may be useful to group the data, if this makes sense. Remember that your graph or graphs must be understandable by anyone unfamiliar with the survey. After you have drawn your graph, write a short paragraph interpreting the results. What story does your graph tell?

3-10 Sketch an appropriate graphic for the following data representing the flight of a flock of hummingbirds from New Hampshire to Columbia, South America. The estimated number of birds leaving New Hampshire during October 2005 was 200,000. The bird counts along the way were the following:

Location	Estimated size of flock
Newark, New Jersey	180,000
Newport News, Virginia	170,000
Wilmington, North Carolina	155,000
Savannah, Georgia	140,000
Key West, Florida	125,000
Columbia	43,000

3-11 The velocity of flow and the dissolved oxygen (DO) were measured in a river downstream from a point of contamination. The data are as follows:

Distance downstream (km)	Velocity (m/s)	DO (mg/L)
0	0.3	0.1
0.5	0.2	0.0
1.0	0.6	0.5
1.5	0.8	1.0
2.0	0.5	3.4
2.5	0.5	3.8

There are at least two ways of expressing these results. Show both graphs. Which graph would be better at demonstrating that rivers will, given time, reoxygenate themselves (the dissolved oxygen will increase)? [Note: mg/L = milligrams of oxygen per liter of water]

CHAPTER 4

PUBLIC SPEAKING FOR ENGINEERS

Public speaking can be a scary experience. The speaker becomes the center of attention and is vulnerable to mistakes, blunders, and other unforeseen disasters. It is no wonder than many people shy away from giving public presentations.

Quite possibly, some people may never have to speak publicly, but not so for engineers. Engineering is a people-serving profession and engineers must be able to orally convey information and ideas to their colleagues, their clients, and the public.

Good speakers make contact with the audience and encourage them to listen. Maintaining contact is not easy. Experience has shown that 30 minutes into a typical college lecture, fully 85% of the students are no longer listening! (Most professors of course deny that this applies to *their* classes.)

So what makes a speech memorable? What keeps the audience on the edge of their seats?

Preparation. There is no substitute for it. Some of the most notable (and notorious) orators in history have carefully orchestrated their speeches and practiced their deliveries. Adolf Hitler, for example, orchestrated all of his speeches, down to the minute gestures and eye movements. Only a few public figures, such as Abraham Lincoln, had the skill and stature to scribble a speech on the back of an envelope and have it become a historic document.

Prepare your speeches by practicing in front of a mirror

108

or by videotaping yourself. These exercises can be ego-crushing, but are well worth the pain and they will help you become a poised and confident speaker.

Below are some other suggestions for making better speeches. One word of warning before plunging in: The best way to improve the quality of your public presentations is to make presentations in public. If this prospect scares you, start small. Speak out in classes, for example. Become involved in an organization that requires you to speak in front of the group (of friends). Seek out a public speaking course. Or, best of all, join a group such as Toastmasters (www.toastmasters.org). You'll be surprised at how much fun it is and how you can learn to *enjoy* public speaking.

4.1 ORAL PRESENTATIONS

Here are some survival tips for public speaking:

Know your audience. What is their level of professional sophistication? How homogeneous is the group? One of the most difficult audiences is a heterogeneous group that requires the speaker to be sufficiently elementary for some people without boring the rest. If the audience has widely varying levels of background, start with simple concepts so that during the first part of the talk you have the entire audience with you but you won't have had time to bore the more knowledgeable listeners. Then move to the more complex material, knowing that you will lose some of the audience but that this will interest others who have been waiting for the more substantive stuff.

Prepare the speech to be too short. Once on your feet, time contracts. You will be amazed that the speech you thought would take 15 minutes has stretched into 45.

Speak louder than you think you should. Vary the pitch and volume of your voice. Remember that emphasis can be added by speaking louder or softer. Great choirs can have their audiences holding their collective breaths by whispering.

Speak more slowly than you think you should. Nervousness causes speakers to speed up. Consciously slow down.

Leave your belongings alone. Do not jingle coins in your pocket, do not twirl your necklace, do not adjust your hair, do not play with your scarf or necktie, do not continuously check to see if your fly is up, etc. etc.[*]

Look good. Don't overdress, but be neat. Remember that you are the most important visual aid in your speech. You are what people will be looking at, and appropriate appearance and effective body language are important elements of any speech.

Stand still if you are behind a podium in a formal setting. In less formal talks you can wander around and

---

[*] Some years ago, when public smoking was still tolerated, I was listening to a magazine editor giving an after-dinner talk when he decided to have a cigarette (during his talk). He started to search around for his pack of cigarettes (all while he was still speaking to us). He finally found them. Slowly he took out one cigarette and continued speaking while holding the cigarette in his hand. Would he put it in his mouth? He did, and then started to search for matches, patting down every pocket in his coat and trousers. Will he find them? He finally did, and then took out a match. Will he light it? He did, but he was so intent on his talk that he forgot to light the cigarette, burning his fingers. New match. Will he get the cigarette lit this time? When he finally lit the cigarette, the audience uttered an audible sigh of relief. Now, years later, I have no idea what the speaker said, but I will always remember his incredibly bad manners and disdain for his audience.

use the space for effect. You can, for example, cross the stage when you introduce a new topic or come up close to the audience when you want to make a special point or respond to a question. (Watch televangelists on television sometime to see how they move to great effect.)

Don't droop or lounge over the podium. It is tempting to use the podium for support. Don't slouch.

Stand tall. A good trick is to stand up on your toes.

Never apologize. Never tell the audience that you haven't had the time to prepare... or that you have a cold... or that your slides are lousy... or whatever. If the slides are truly lousy, the audience will know soon enough. There is no sense in emphasizing an already deplorable situation.

Be careful of jokes. A quick and dirty way to die in front of an audience is to tell a joke that isn't funny.

Finish your sentences. Some speakers have so much information in their heads that they simply overwhelm themselves and start innumerable sentences, never finishing any. Words need to be in complete sentences in order to make sense.

Don't talk to yourself. Many speakers think out loud, such as "Let's see now ...?" or "This is not straight on the screen," or "Where do I go from here?" Such musings are highly distracting and give the impression that the speaker is unprepared.

Be careful of the magical words "In summary...." or "In conclusion... ." You will raise expectations that you may not wish to fulfill. Only say those words when indeed you are about to conclude the speech.

Have a strong beginning and ending. In a public presentation, remember that your goal in the first few minutes of a speech is to convince the audience that you are worth listening to. You have their undivided attention for the first two minutes so make the most of it. In speeches, as in writing, the first and last sentences are the most important. The first sentence is in the primacy position, and gives the audience the first, most important impression of you. The last sentence is in the recency position, and is what people remember the most. Make the beginning and the ending truly memorable.

Don't start by thanking your audience and don't end by thanking your audience. Get them in the beginning with a zinger of a first sentence and leave them with a strong conclusion that clearly signals the end of the speech. Saying "Thank You" is like saying "Amen."

Decide whether it is better to read your speech or to use notes. Whatever you choose, the method should correspond to the occasion. For example, when the President of the United States delivers his (or her?) State of the Union address, she or he invariably has it written out and each word is carefully considered. In contrast, a press conference is a more *ad lib* affair where the President answers tough questions without the benefit of notes.

If it is better, for whatever reason, to speak from a written text, the delivered speech must sound like conversational speech. This can be accomplished by first dividing the text into phrases from one to ten syllables long with slash marks at appropriate places. The slash marks create *intonational units*[1] containing an appropriate number of syllables that make the reading sound like conversational speech.

For example, read the following sentence out loud, just as it is written:

The slash marks create intonational units containing an appropriate number of syllables that make the reading sound like conversational speech.

Now read out loud the sentence below, pausing at every slash mark as indicated.

The slash marks / create intonational units / containing an appropriate number of syllables / that make the reading sound like conversational speech.

The second way of reading out loud should sound much better. The slash marks force the speaker to pause at approximately the places where he or she would have normally paused if speaking extemporaneously.

Use simple sentence construction in your speech. In informal conversation people seldom use long words or convoluted grammar, and the same should be true in a speech. Abraham Lincoln got away with it in "Four score and seven... ," as did Franklin Roosevelt with "This day will live in infamy," but these were speeches for special occasions. In everyday public speaking, use small words.

Try to develop a rhythm in your speech. The rhythm of a speech can greatly enhance its effectiveness. For a perfect example of such a speech, listen to how Martin Luther King, Jr. used rhythm to perfection in his "I have a dream" speech during the march on Washington at the height of the civil rights movement. A speech uses rhythm and repetition for emphasis. A good speech should start to flow on its own, with the listeners in rapt attention, almost writing the speech in their minds as the speech progresses. Ahmad Jamal, an outstanding jazz pianist, is famous for leaving large chunks of silence in his performance, allowing the audience to fill in the missing notes. A good speaker can use rhythm and spacing to create the same atmosphere in a speech.

Use silence. Silence is a powerful tool, so use it. If you want to make a special point, look at the audience and say your piece, and then pause. Let it sink in.

Use appropriate gestures. Some speakers develop bad habits that are painfully obvious to the audience but totally oblivious to the speaker. For example, some speakers will use the "Truman chop", named after President Harry S. Truman who used it (probably unconsciously) in his speeches. The Truman chop resembles the "Atlanta Braves Chop" or the "Florida State Chop." There is nothing wrong with this gesture, or any other gesture, if it is used in moderation and for appropriate emphasis, but over use can be a distraction. The best way to improve your own use of gestures is to videotape a talk you give and then study it later. A speech with **no** gestures, by the way, can be dull and dry. Use gestures to bring the speech to life.

Control the question and answer period. Most public presentations have a question and answer period, and this can be scary if you do not control the tone and pace. When you get a question from the audience, first make sure the audience has heard the question. A useful technique is to repeat the question, thus allowing you to rephrase it and buy some time to formulate an answer. Do not engage in prolonged arguments with sarcastic or antagonistic questioners who are just trying to show off. If you do not know the answer to a question, say so. If the question is irrelevant, or if the answer would take much too much time, offer to meet the questioner after the formal presentation.

4.2 VISUAL AIDS

4.2.1 The value of visual aids

Psychological studies have repeatedly shown that information is transferred best if two or more senses are used simultaneously. About 35% of all the information we retain is visual, while only about 15% is verbal. But if the two are combined, the effect is synergistic. For proof, look at any advertisement on television. Very seldom will a significant word be spoken without it also being printed on the screen.

Effective use of visual aids can make or break a presentation. The types of visual aids used in public speaking vary widely – from material demonstrations, to overhead charts, to videos. The most important rule for visual aids is that they must all be necessary for the speech. A public presentation is not an opportunity to display a scrap book of neat looking but irrelevant pictures.

The single most important issue in the use of visual aids in public speaking is that the speech must stand by itself; the visuals simply make it better. Some speakers have all the information on PowerPoint slides and begin the speech with the words "Let's just move to the first slide" and they start to read the slides. This is an insult to the people in the audience who can read the slides very well by themselves, thank you. The purpose of the slides is not to provide you with the notes for your speech.

Another potential problem with using slides for your entire speech is that if, for whatever reason, the technology bug bites and your lovingly prepared slide presentation cannot be projected, all is not lost. Your speech should stand on its own and the slides should only improve it.

4.2.2 The special case of PowerPoint®

Years ago it seemed that nobody in business or management could give a talk without overhead transparencies. The overhead projector sitting in the middle of a tiered classroom became the icon for business schools and corporate board rooms. Nervous speakers would simply read the slides to the audience.

This situation changed dramatically and suddenly with one incident that sent shudders through the business world. In the early 1990s, the new president of IBM, Louis Gerstner, was traveling around getting acquainted with the company, and at every stop he was subjected to overhead slide presentations. It seemed nobody could give a talk without overheads. Finally at one meeting, as the speaker was about to launch into yet another overhead slide presentation, Gerstner walked up to the front of the room, turned off the projector, and asked the flabbergasted speaker to just **talk** about his projects. The incident sent e-mail shivers throughout IBM, and the grip of transparencies on business presentations was broken.[2]

The same situation seems to be developing in engineering, but in our case the technology is Microsoft's PowerPoint®. It seems that today any engineering student who does not use PowerPoint is simply not with it.

PowerPoint technology is clever, simple, and easy to use. Better projectors can now show the images without having to darken the room, and changes to the slides can be made quickly and seamlessly. It is the ideal technology for engineering presentations, and thus PowerPoint has become a ubiquitous part of engineering. For this reason, it is proper to devote some time and space in this book to PowerPoint, even though it is a proprietary visual aid.

There is no need to talk here about how to prepare a PowerPoint presentation because every engineering student should know this already, and if not, there are hundreds of

116

tutorials on the web that can help. Instead, let's concentrate on how to use it **properly.** And our aim is not to dis PowerPoint as seems to be the fashion in many articles and books. PowerPoint is a useful tool, and we engineers need to learn how to use it well. Or, in Mark Anthony's words (sort of): "I come to praise PowerPoint, not to bury it."

For starters, let's suppose you saw the following PowerPoint slides, but did not hear the presentation for which they were prepared. What do you think was the subject of the presentation?

New product
"Rocket"
Fruit
Mexico - New York

Low interest
Accelerated process
Expertise available

Cost acceptable
Cayman Islands
Return

You might properly think that this is what is being talked about:

Slide 1
The company is getting ready to enter the aerospace field and is going to design and construct a new rocket-powered aircraft. The main objective of this aircraft, and the most likely market for it, will be the tropical fruit companies that have to fly the fruit grown in Mexico to the processing plants in New York and do this as quickly as possible.

Slide 2
The schedule for the design of the aircraft will be accelerated to take advantage of the low interest rates and the surplus of technical workers created by the economic slowdown.

Slide 3
The cost of the project is expected to be high, but the money can be borrowed from banks in the Cayman Islands. The return on investment is expected to be substantial.

But you might be wrong. Instead, the speaker may have been talking about this:

Slide 1
The company is getting ready to introduce a new line of foods for young children. The plan is to start a new product line called the "Rocket" that will be produced in the processing plant in Mexico and in New York. The product will include many fruit flavors.

Slide 2
Children of that age have little interest in baby foods, and parents want to speed up the mealtimes. Surveys have shown that the "Rocket" foods taste better and therefore the

eating process can be accelerated. Parents already have the expertise to do this, but a new better-tasting food is needed.

Slide 3
The cost to the parents is acceptable even though the new "Rocket" food will be higher priced than Gerber's. Studies done in the Cayman Islands have shown that the number of babies who barf up the new baby food is much lower than for normal baby foods.

Silly? Yes. But the point is this: A lot is lost when information is condensed to only key words, and this is the greatest danger of PowerPoint. The process of presenting information on slides can result in a loss of information. When sentences and qualifiers and modifiers are eliminated, the meaning is often lost or modified.

The objective is therefore to use PowerPoint in such a way as to **enhance** the presentation and not to replace it. The greatest mistake engineers (and others) make in the use of PowerPoint is to believe that all they have to do is to prepare the slides and then simply read them to the audience. This is not only dangerous, it is also dreadfully boring.

Here are some other recommendations on the use of PowerPoint in engineering presentations:

Consider not having a title slide. This is heresy for some engineers. Every engineering presentation you have ever seen probably began with a title slide that included the title of the talk, the name and affiliation(s) of the author(s), and other stuff, including advertisement. (Incidentally, some companies have a PowerPoint slide template that has the company logo in the corner and everyone is expected to use this. Some people find this incredibly annoying and in poor taste, especially if the presentation is supposed to be non-commercial.)

But why eliminate the title slide? The reason is focus. After the speaker is introduced, the focus of the audience ought to be on the **speaker.** If a title slide is shown immediately after (or sometimes before) the introduction, the audience will start reading the slide and not concentrate on the person who is speaking to them. In the beginning of the presentation it should be all about you, the speaker, and not about the subject of the talk. After you are introduced, you thank the introducer for the introduction; you look out over the audience and make eye contact; you say something friendly and/or humorous; and you generally ease into your presentation. After having established this contact with the audience, it is now safe to focus their interest away from you and to the slides.

There is, however, a good argument for having a title slide if you are presenting a paper at a conference where the majority of the audience might not know you. In such cases having your name and affiliation in front of the audience at the start of the presentation is useful. So there are times when a title slide might be appropriate, but think about its use and purpose and don't just automatically use a title slide if it makes no sense to do so.

The best way (personal opinion here!) to start a presentation is to use a blank slide. This way the projector can be on (you know it works and it does not have to warm up) and the first slide is only a click away.

Consider not having an overview slide. The second icon of all engineering talks is the overview slide. But why is it necessary? The audience knows full well that the first thing you are going to do is to have an introduction. They also know that you will have a conclusion. They also have a pretty good idea of what is in the middle, and if they don't, they will be pleasantly surprised. So eliminate the overview slide. It is unnecessary; it takes time away from important things; and in some ways it is an insult to the audience.

You don't have to point to everything on the screen.
Your audience is intelligent enough to be able to read what
is on the screen, or to appreciate the picture you are
showing. Do not insult the audience by indicating every
word with a pointer. Pointers are useful (and even
necessary) when there is a more complex picture that needs
explanation. For example, if a map of a region is projected,
it might be useful to point out familiar places on the map in
order to orient the viewer.

Stand at an appropriate spot when using slides. First,
and most obviously, do not stand in front of the screen so
you block the view. If you are in a small room and are
going to point to items on the slides with your hand or a
pointer stick, stand to the left of the screen (as the audience
sees it) and use your left hand. If this is uncomfortable,
stand to the right of the screen and point to the slides with
your right hand. Never turn your back to the audience when
pointing to the screen!

Be careful with a laser pointer. Laser pointers are
excellent tools, but their use often is marred by two
mistakes. First, there is the temptation to face the screen
and point to items on the screen, again turning your back to
the audience. Avoid this at all costs. Always … face … the
… audience. The second mistake is that some speakers
don't turn off the pointer when they have used it to point to
the screen, and sometimes the laser beam even dances
around the audience causing great distress as people are
ducking their faces and lowering their eyes to avoid contact
with the beam. The proper use of this excellent tool is to
point to the screen as needed, and then turn it off.

Do not read the slides. Do not read the slides. Do not read
the slides. Do not read the slides. Look… at… the…
audience!

Suppress your desire to be clever. PowerPoint is a powerful tool. It has bells and whistles that most of us will never use, and occasionally it is fun to see what a true PowerPoint jockey can do with it, but most of the time such stuff is distracting. Don't try to overwhelm your audience with how clever you are at using PowerPoint. This is not the objective of your talk and doing so will distract from your message.

Be careful with visual humor. Don't try for humor by putting a risqué slide in, and then saying "Oops! How did *that* get in there?" There is no easier way to alienate your audience.

Trust that the slide is still on the screen. Too many speakers keep looking back at the screen, apparently to check if the slide is still there. It is. No need to check. Eliminate the nervous habit of looking back. Look at the audience. Look… at… the… audience! Keep them on your team and interested in what you have to **say**.

Subordinate the visuals to the speech. Use the slides to your advantage, but do not make them the notes from which you speak. If you put your outline on the screen, then you are showing the audience your working draft. It's kinda like showing your underwear. It's just bad manners.

A pragmatic reason for not using PowerPoint for your notes is that you would be a dead duck if for some reason the technology fails. Your speech should always be independent of your visuals.

Finally, you should not use PowerPoint in place of notes, because you will lose degrees of freedom. You might, in the middle of your presentation, want to add something, or tell a story, or respond to a question. If your notes are on the slides you will not be able to do so without messing up your continuity. *

Check and recheck. Run through the PowerPoint slides before the presentation in order to make doubly sure they are in the correct order and that you don't have any misspelled words or other goof-ups.

Make the writing or graphics on your visuals "too big." There are rules about font size but it is not necessary or useful to specify these. The best way to make sure that your font is the proper size is to go to the room where the presentation is to take place and run through the slides making sure that the people in the very back row will be able to see them.

Be careful about using visuals made from printed material. Visual aids for public oral presentations should be different from figures used in publications. Illustrations used in oral presentations should be open, clear, simple, colorful, and big. The listener does not have time to scrutinize the figure in order to understand it, so the figure must convey the argument quickly. Horizontal lettering is better than vertical lettering because the audience cannot turn the paper sideways to read the vertical words. Complex pictures and graphs scanned in from printed publications are especially susceptible to this problem. A reader of printed material has time to figure out the graphic,

---

* As previously noted, one of greatest speeches of all time was Martin Luther King, Jr.'s "I have a dream" speech in Washington DC during a civil rights rally. But it turns out that he never wrote that speech. He was in the middle of his prepared remarks when he felt the audience leaving him, and he decided on the spur of the moment to change his speech. If you have ever seen videos of it, you will see and feel how the audience came to him and responded to his words. It gives you goose bumps, and it would never have happened if Marking Luther King, Jr. had used PowerPoint to project his outline.

but the audience at an oral presentation might not have time to understand the message.

Bullets. The multiplication of bullets over the past few decades has been astounding. It is difficult to believe that Shakespeare, for example, did not use a single bullet and still seemed to be able to write some pretty good stuff. Bullets can be useful, but their proper use must be understood.

Recall from Chapter 1 the different kinds of lists. A numbered list implies sequence, and this would be appropriately used for instructions. A lettered list can also imply sequence, but can also be used to imply importance, with the first item listed, under "a", being the most important and the importance diminishing down the list. Both numbers and letters can also be used for identification purposes. For example, the problems at the end of the chapters in this book often have lists, and the items are identified by letter or number for instructional reasons.

PowerPoint, however, is the home of bullets. As noted previously, a bulleted list implies that all of the items in the list deserve equal weight, and this is the requirement that is violated most often. For example, suppose your company is trying to decide where to open a new sales office, and you have a slide that looks like this:

- **Pittsburgh**
- **Steelers**
- **Ketchup**
- **Cleveland**
- **Lakefront**
- **Airport**

You are trying to say that the advantages of establishing the office in Pittsburgh is that one can go to the Steeler games, and that Heinz makes great ketchup. The advantages of Cleveland are its waterfront and the excellent airport. But the descriptions of the advantages are clearly subsidiary to the cities, so the slide ought to look like this:

- **Pittsburgh**
  - o Steelers
  - o Ketchup
- **Cleveland**
  - oLakefront
  - oAirport

Project your slides as needed, and then remove them. Remember that visual aids should be subordinate to your talk and that when you are not using them, you want the audience to focus on you. When you are finished with a series of slides, insert a blank slide. You want the audience to focus on you again, and they will not do so as long as there is a picture to look at.

Anticipate the next slide. You should either memorize the sequence of slides, or if you are like me and need a cheat-sheet, show in your written notes where the slides are. Then as you are talking, you know what the next slide is and you can begin to say something about it as you simultaneously show it on a screen. This makes for a smooth transition from slide to slide and gives the talk a sense of preparedness and continuity.

Have a strong concluding slide, and then remove it. After you have stated your conclusion and shown it on the screen, click to a blank slide or blank screen (you can hit B on the keyboard and this turns off the slides). Do not leave your last slide on the screen during the discussion period. It is best to leave the projector turned on, however, in case there are questions that require you to go back to a slide (you can hit B again to get back to your slides). When you are finished with your presentation, you want the audience to focus on you the speaker so that you can respond to questions. A cute slide with "Questions?" on it is not very professional and should be avoided.

4.3 LISTENING TO SPEAKERS

Just as effective public speaking is a skill that can be learned, so is listening to speakers who are trying to tell you something.

Perhaps the most important listening skill, especially for students, is note taking. The ability to take notes in a class can often be the difference between success and failure. How is it that some students can take wonderful notes that read like an outline of the lecture, while other students cannot seem to scribble anything comprehensible? The answer is that some students have developed their listening skills while others have not.

This skill is also useful in engineering practice because engineers often need to gather information from many sources in order to solve engineering problems.

Here are some suggestions for good listening:

Get enough sleep during the night so you stay awake during a presentation such as a lecture or seminar. You might be surprised that this admonition appears in an engineering textbook, but experience has shown that too many students are trying to cram entirely too many hours into the available 24. Take time to sleep. You will be amazed how this will help your listening skills, and your grades.

Be alert to significant points. During the lecture or another presentation, listen for such clues as "Now this is really important," or "What does all this mean?" The speaker is about to present the bottom line – the one piece of information he or she really wants you to understand and remember.

Organize your note paper. One effective way to organize your note paper is by dividing it vertically into two halves. On the left side write what the speaker seems to be saying. After the lecture, on the right side rewrite the important points presented during the lecture. At the conclusion of the lecture draw a horizontal line at the bottom of the notes and write the questions that should be answered in the next class.

Before the next class, read over your notes. The professor will simply assume that you remember everything from the previous class and will go on with the topic.

Maybe not even take notes? In some classes, note taking may not even be worthwhile. This is particularly true in science and mathematics where the professors try to explain theorems or principles. This material is no doubt in your textbook and is written with much more clarity than anything you can copy from the board. Concentrate instead on *understanding* the discussion, and ask questions whenever you get lost in the argument. Do not allow the professor to proceed unless you understand the material to that point. You will be surprised how many professors will appreciate this, because they seldom receive immediate feedback and often do not know where they lose the class.

Ask clarifying questions. Asking questions in class or at a professional meeting is risky and not everyone will participate. Asking a question that everyone else knows the answer to will hang you out there to dry, slowly twisting in the wind. (How is that for two metaphors back-to-back?) In short, many people don't want to risk being embarrassed and therefore they will not ask clarifying questions. The truth is that the very question you have in your mind is probably shared by the rest of the audience, and every person there wants to ask the question but is afraid to. Most speakers (especially professors) are pleased to get questions, and often revel in them, going off on interesting tangents. So hitch up your courage and ask.

REFERENCES

1. The idea of intonational units is credited to Chuck Larkin, in *The Storyteller-Entertainer Handbook*, copyright by Chuck Larkin, printed by G. K. Thompson, Atlanta, GA, 1984.
2. Gerstner, L. V., Jr., "Who Says Elephants Can't Dance? Inside IBM's Historic Turnaround," Harper/Collins, New York, 2002.

PROBLEMS

4-1 Critically observe a public speaker. This could be a professor, a televangelist, a politician, or any other person making a public presentation. Focus more on the style of the speech than the content. Pay attention to all the elements of good public speaking and watch for mistakes. Take notes on the speaking style and write your criticism as a one page (single spaced) memorandum.

4-2 Choose one oral presentation you have had to make. Describe the audience and the purpose of the talk. What went right and what went wrong? How could you have improved it? Report as a one page memorandum.

4-3 Choose one professor and analyze the professor's method of responding to questions. Are the students encouraged to ask questions or is the atmosphere intimidating? Does the professor rephrase difficult questions and try to get to the heart of the matter? Is there ever an admission of not knowing the answer to a question? Write your analysis as a one page memorandum to the dean. (Do not name the professor!)

4-4 Write down five topics on which you would be able to make a two minute (maximum!) speech at a moment's notice. These will be collected by the instructor, and you may be asked to make such a presentation at any time. At least three of these should have something to do with engineering.

4-5 Choose one particularly bad class you are trying to survive and try the technique of note taking described above – dividing your note paper in half. In response to this problem, report as a one page memorandum on the advantages and disadvantages of the strategy. Would you use it again and why or why not? Include as an attachment a copy of your class notes.

4-6 Watch a White House or other Washington press conference and note how the speaker either avoids or responds to questions. What questions would you have wanted to ask?

4-7 Watch a televangelist do his/her stuff and pay special attention to the gestures. How do they use gestures for emphasis? Are they natural or do they look contrived? Do they use body language in any way, and is this effective? Write a short description of what you observed.

CHAPTER 5

ENGINEERING CONVERSATIONS

"Public speaking" is an act of speaking professionally to a group of people – making a public presentation. Another professional oral communication is called "conversation," or speaking one-on-one. Conversations differ from public speeches in that in conversations there is a two-way discussion. The speaker is also a listener and reacts to what the other person is saying.

The conversations discussed in this book occur within the context of the profession. The speaker would be speaking as an engineer, in a professional capacity, and would have all the responsibilities and privileges of that profession.

In this chapter we cover four types of conversations you will undoubtedly have during your engineering career. First, it is quite likely that you will have an interview for a prospective job, and a very important part of that interview will be a personal discussion with the potential employer. Once employed, you will sooner or later have to converse with clients and vendors. If you are employed in the public sector, then you will also have to talk to representatives of the media and to attorneys. Most practicing engineers are quite uncomfortable in these circumstances because they do not know the rules that govern such conversations. The discussion below provides some helpful hints on how to survive conversations with prospective employers, clients, reporters, and lawyers.

5.1 JOB INTERVIEWS

If you are applying for an engineering job with a private firm or a governmental agency the interview will be conducted by either an engineer or by a human resources person. If it is an engineer, he or she may not have the foggiest notion of how to conduct an effective interview and you might have to take charge of the conversation. In most cases, however, you will have a conversation with someone who has done this before, and the interview will be a pleasant experience.

Pleasant or not, here are some helpful hints for effective job interviewing:

Wear nice clothes but do not overdress. For men, a sport coat and dress shirt, with a subdued (not flashy) tie would be appropriate. Don't overdress. Engineers are pretty informal folks, and you don't want to stand out like a Brooks Brothers model. But also beware of being too casually dressed. You are, after all, on display and you don't want to look like a slob. Get a haircut (or at least comb your hair), clean your fingernails, and do all the things your mother told you to do.

For women, the same general advice applies. You want to be comfortable but not informal. Dress to be taken seriously. Stiletto heels and low-cut sweaters will only distract. A business suit is appropriate. Sandals and tank top would be unacceptable.

Since a sense of appropriateness of dress is not something one can put on a resume, the interview is an opportunity for the prospective employer to see if you know how to dress appropriately.

The introduction is critical. When you are introduced, be confident but not arrogant. Shake hands with a firm handshake and look the interviewer in the eye.*

Get involved. Once the interview begins, allow the interviewer to ask the first question, but then become actively involved. If your interviewer brings up a topic not related to the job, and if you have any knowledge of this topic, jump on it. Get the interviewer to start talking about things he or she likes, and establish a bond on a non-job level. You need to convince the interviewer that you are the kind of person he or she wants to work with and talking about non-professional matters is often a great way to bond with the interviewer.

Be prepared. Read the company or agency literature and know something about how the outfit operates. You don't want to ask inane questions that you ought to have found out before the interview. Also be prepared for stock questions that interviewers like to ask to see how the candidates respond. Anticipate questions such as:

- Why would you want to work for this company?
- How do you see your career developing?
- Do you want to get into management or are you set on staying in engineering?

and the most asked question of all,

- Do you have any questions?

Be prepared for this question, and have something

---

* This advice is obviously culturally specific. An interview in Japan, for example, would have its own very specific rules, including the exchange of cards and bowing. In other countries looking a person in the eye when shaking hands is considered bad manners. If interviewing outside the United States, do some research to find out what the proper customs are.

interesting to ask.

Be honest. Don't inappropriately embellish your skills or experience. Speak with candor.

Do not volunteer negative information. Be positive about your experience and skills. Imagine the impression created by a candidate volunteering: "I really did not like the mechanics sequence. I don't think I was right for it," or "I want to get into sales, but I am not sure I have the confidence to do it," or "Well, of course I put the best references in my resume. I wouldn't want you to talk to some people who would not have much good to say about me!"

Some questions are out of bounds and you should not answer them. Questions that may lead to illegal discrimination on the basis of race, sex, religion, national origin, or physical disability are prohibited by law. For example, "Are you married?" "What language did you first learn to speak?" or even "Have you ever run in a marathon?" You can volunteer such information if you believe it is to your advantage, but it is not legal for the interviewer to ask for information that does not directly relate to the job.

These questions are illegal because your answers might result in your being discriminated against on the basis of characteristics that have little to do with your job, although how these relate to possible discrimination is sometimes unclear. The law assumes the worst possible situation and the most prejudiced of interviewers. In theory, if the interviewer has deep prejudice against some ethnic group, for example, then admitting that this was your first language might lead the interviewer to have a lower opinion of you. Such questions are therefore not allowed.

Ask to talk to other people. Sometimes, if you are lucky, you will have a chance to talk to other engineers on an informal basis. This is always a good chance to ask tough questions. These people are your kin, of course. You are one of them. They see themselves in your position in the not too distant past and will be honest with you. Ask about the general tenor of the work place, overtime expectations, and opportunities for professional development.

Take the interview seriously. It goes without saying that you should take the interviews seriously. Don't be flip or show off or do something unprofessional. This admonition is not trivial. A private corporation recently asked its interviewers to relate the most unusual incidents they had personally encountered, and here are some of the responses:[1]

• The reason the candidate was taking so long to respond to a question become apparent when he began to snore.
• "Why go to college?" "To party and socialize."
• Said she had graduated cum laude but had no idea what cum laude meant. However she said she was proud of her grade-point average of 2.1.
• She actually showed up to an interview during the summer wearing a bathing suit. Said she didn't think I would mind.
• I had asked the candidate to bring a résumé and a couple of references. He arrived with the résumé – and two people.
• Without asking, he casually lit a cigar and then tossed the match onto my carpet – and couldn't understand why I was upset.
• The interviewee had arranged for a pizza to be delivered to my office. I had to ask him not to eat it until later.

5.2 SPEAKING WITH JOURNALISTS

Journalists walk a narrow line between professionalism and sensationalism. Some journalists are professionals, seeing themselves as a link between the public and the world, including government, science, entertainment, and society in general. They vehemently and eloquently defend the journalistic enterprise as a high calling, are careful and fair in what they print or what they say, and recognize the importance of their role in upholding our first amendment rights.

But journalism also has a less professional side. When the pressures of deadlines and money dictate journalistic activities, even some of the most well-regarded news organizations can stumble and place expediency ahead of professionalism. All news organizations (in a free society) recognize that if nobody bought the papers, and if nobody watched the tube, then there would be no newspapers or television.

Such pressures are a constant concern for even the most modest hometown papers and radio and television stations. Most of the reporters whom engineers encounter work for the local press, and it is on the local level where engineers are most visible to the public and where engineering most often influences people's everyday lives. Unfortunately, engineers often do not understand that the role of a reporter is very different from that of a professional engineer and these engineers treat reporters as they would their professional engineering colleagues. Many engineers are unaware of the pressure on reporters to publish what editors demand or whatever sells more newspapers or commercials.

Remember that local reporters for both the print and electronic media work for a living and that they get paid for obtaining newsworthy material. While most reporters are interested in relating a balanced account of a news story, some will also try to get the principals (often an engineer) to say something that will result in headlines or sensational soundbites. At times the engineer can use the reporter to further his or her own objectives (such as preparing for a public hearing) but other times the reporter can get the engineer to say something that is not in the engineer's best interest. The reporter views this as simply doing a good job and usually there is no malicious intent.

Reporters admit that during an interview the interviewee (the engineer) is in the power position. The journalist will do everything possible to reverse those roles and it is up to the engineer to prevent the switch.

Here are some pointers on how to prevent losing your edge to a reporter:

Do your homework. If you have time, find out what the reporter really wants, what he or she has written on this topic prior to your interview, and what he or she wants to write this time. If you are working for a company or any organization, other than a university, you need to get permission from your superiors to be able to talk substantively to the media. Finally, decide ahead of time what you want to talk about and stick to your agenda.

Try to understand what the reporter's perspective is and respond accordingly. In general, the media distinguishes factual reporting from editorializing, but personal opinion creeps into ostensibly factual news stories. Every person, reporters included, has a bias based on background or education and this may be quite evident in the reporting. Even the decision to do a story is a value judgment, because this defines what is and is not news. Before starting the formal interview, try to understand the reporter's perspective. For example, the reporter may have heard only one side of the story and has made up his or her mind as to what the facts are. In that situation it may be necessary to emphasize the opposing view. Or it may be that the reporter has a strong anti-environmental bias and is hoping that the engineer will say something embarrassing. In any case, try to interview the reporter first and establish why it is that you are being interviewed.

Do not feel pressured by time or the presence of a camera. The reporter's deadline is not your problem. If you get flustered, you are much more likely to say something ill-considered or misleading, and this could result in an inaccurate story.

If you need more time to compose an answer, ask the reporter to repeat the question. Remember that the reporter needs you, not the other way around.

Once you have answered a question, ask the reporter to explain the answer back to you to see if he or she has understood. Often reporters are apt to ignore answers, thinking instead of the next question they want to ask. Asking the reporter questions can help clarify the point and prevent inaccurate reporting.

If the reporter works for the print media request that he or she call you back and read the entire story to you to make sure the facts are correct. Often they will not do that, citing deadline pressures, but it's worth asking.

Set the ground rules before the interview begins. There are three modes of interviewing: *on the record, off the record,* and *background.* If you agree to be on the record then everything you say can be used and attributed to you directly. If you insist on being off the record then nothing you say can be used directly. ("A senior White House aide confirmed that...") Background means that it is not permissible to even acknowledge that the conversation has taken place. The purpose of a background interview is to educate the reporters so that they can be better reports.

If you are not the best person to answer a question, suggest that the reporter ask the person who is. Do not speculate on how other people might respond to a question.

Use eye contact wisely. Eye contact is important during a press conference with several reporters. If you hold eye contact with a reporter, you are in effect asking for a question. If a reporter asks a hostile question, answer it and look away.

Reframe questions as appropriate. Sometimes it is advantageous to repeat the question, but reframe it in the way you want to answer it. The same technique can be used during question and answer sessions at technical meetings, as discussed previously. [*]

Be terse on camera. In replying to TV reporters, remember that your answer must be in the first four or five words. That is all that will be used in the news program. Try to remember throughout the interview that every sentence and short phrase can be clipped and used in a "sound bite" or quote.

---

[*] When President John Kennedy appointed his brother Bobby Kennedy to the post of Attorney General a lot of people had concerns about having two brothers in such powerful positions. At a press conference President Kennedy was asked, "Do you not think it is improper to appoint your brother the Attorney General?" He responded by restating the question as he wanted it, "The question relates to Bobby's qualifications to become the Attorney General," and he then proceeded to list his qualifications, completely ignoring the thrust of the original question. He should, of course, have said "No, I don't think it is improper," and the purpose of this example is not to suggest that such political weaseling is necessarily proper. It is, however, a classic example illustrating how a difficult question can be deflected.

It's fine to be silent. Remember that you are in control and that silence is part of any conversation. Do not start to babble just to fill a void.

Do not say "No comment." Explain why you cannot discuss something. Otherwise, people will suspect the worst.

Answer positively. Do not speak negatively of another person. You have little control over what will actually be printed or aired, and you could make someone very angry without having the opportunity to explain.

Be honest. Most importantly, be honest. Reporters are trained to sniff out obfuscation and misleading statements. If you are not sure of something, say so. Don't try to make something up to enhance your image.

5.3 SPEAKING WITH CLIENTS

Engineers are often called on to speak with clients about the work the engineering firm is undertaking. The engineer has technical know-how that is being bought by the client (or else the client would not hire the engineer) but the client is the one with the money that the engineer desires. This delicate dance between unequal partners requires diplomatic skill.

Engineers talk with clients, and potential clients, at numerous stages of any engineering project.

The first kind of presentation engineers make to (potential) clients is the "Request for Qualifications." When a potential client has a project that needs done it broadcasts the need for an engineer and then various firms respond indicating interest. From these, the client chooses some it wants to talk to further and asks the firms to submit their qualifications.

Often the client will consists of a committee charged

with making the decision, and the engineer has an opportunity to convince them that the engineer's consulting firm is indeed qualified to do the job. The presentation consists of descriptions of previous jobs like the one being contemplated and some understanding of the client's problem.

Based on these presentations the engineering firms are asked for full proposals, and these are both written and oral. The oral presentations are again in front of the selection committee. Following these presentations the client will select the engineering firm it wants to hire, and a contract is signed, which includes the compensation for the engineering work and the time-line for the project.

After the contract is signed, the engineer is contractually beholden to the client and is responsible for completing the project. But the engineer must also remember that the contract can be voided at any time due to "unresponsiveness" or even "non-performance." In other words, the engineer can be fired at any time, and thus there is constant effort to satisfy the client. This usually takes the form of periodic conversations.

During the time the engineer is working for the client, the engineer is beholden by the Code of Ethics of every engineering discipline to be a "faithful agent." That is, the engineer is working only for the client on matters having to do with this job. There can be no conflicts of interest, and there has to be total transparency. The engineer speaks for the client in public meetings very much like a lawyer speaking for his or her client in a criminal case.

When a project is completed, the client will usually have a public hearing (if the project uses public money) or a meeting of top management (if the project is for a private firm). At this meeting the engineer's job is to convince these groups that the engineer has responded to the needs as originally specified in the contract.

When the engineer in private practice is paid, he or she

is discharged of further direct involvement in the case. That is, the engineer cannot be called back to do more work without a new contract.

During any engineering project the engineer has many conversations with the client during which the engineer presents a progress report or asks the client for guidance on how to proceed. During these meetings the engineer should dress professionally and exhibit a decorum that commands respect. It is fine to be friendly with clients before the meeting, but during the meeting these individuals ought to be treated formally and referred to by their professional or political titles.

> BAD EXAMPLE: "As you can see, the cost will not exceed the budget. Not like your checkbook, eh Joe? Heh, heh. And we won't have any trouble with the planning commission. I understand Mike has them in his pocket."
>
> REVISION: The budget seems to be reasonable and we do not expect cost overruns. Thank you for that question, Mr. LeVan. And concerning the planning commission, I understand that Mayor Nelson is very much interested in this project and will be able to help us with the permits.

5.4 SPEAKING WITH LAWYERS

Engineers, on a professional level, speak with lawyers in three different circumstances: assisting the lawyer in the preparation of a case, providing a deposition, or appearing in court.

When the engineer is assisting the lawyer in the preparation of a case, the engineer and lawyer are cooperating and have a common objective. The engineer is paid a professional fee by the lawyer's client. In most cases, the issue is settled before it goes to trial and the engineer may never have to appear in court or participate further in the case.

Engineers are often asked to provide *depositions* as expert witnesses, in which case they are again working for one side in the dispute. A deposition is a record of questions and answers, recorded by a court stenographer. The transcript becomes part of the legal testimony in the case and is presented by the lawyers in court. This is called the *discovery* phase of the case. All depositions are shared with the opposing side. Depositions save court time by gathering as much factual evidence as possible before the case goes before a judge.

When the case begins, the lawyer who has hired the engineer will go over the finer points of the case and try to anticipate the questions that the opposing lawyer might ask. This "prep" is very important and useful for the engineer. If the lawyer is unwilling or unable to spend time in explaining the case fully, the engineer might consider withdrawing from the case. The pay might be good, but the aggravation of being destroyed on the stand by clever questioning is hardly worth it.

When the engineer assists the lawyer in court, the engineer is an *expert witness* and is allowed to express opinions concerning matters with which the jury might not be familiar. Once again the engineer is working as a consultant for one side in the dispute and receives compensation for his or her time. Ideally, as an expert witness, the engineer has no interest in the outcome of the case and is simply testifying factually.

The opposing counsel, however, views the engineer as an adversary and will attempt to either discredit the engineer, discredit the testimony, or get the engineer so flustered that he or she gets angry and says something inopportune. In any case, this is not a pleasant experience for the engineer who is trained in cooperation and consensus. Engineers often do not understand that law is

142

adversarial, and **the intent of the opposition is not to discover all the facts in the case, but to win.**

This is the basic philosophical difference between the way lawyers work and engineers work. Engineering is usually not adversarial. What is important to engineers is the solution to the problem, and any available information is used to advantage. In law there is a disagreement that must be resolved and this creates two opposing sides, only one of which will win. The lawyer's role is to help his or her client win, and not to seek justice or to establish the truth.

Lawyers are taught to be ruthless in the pursuit of their goal of winning but not to take their work personally. They are sometimes required to represent clients whom they know to be guilty. Engineers, on the other hand, internalize criticism and can't understand how two lawyers can be so nasty to each other in court and then socialize as friends afterward. When engineers are drawn into the legal arena, they must recognize that they are in a different world and must act accordingly. They must remain dispassionate, clear in their testimony, and above all, brutally honest.

REFERENCES

1. McShane, L. "Want a Job? Don't Have Pizza Delivered During the Interview" *The News and Observer* Raleigh NC, 14 December 1997, p. 2E, quoting a study conducted by the Commemorative Brands Company.

PROBLEMS

5-1 Ask permission from a local television reporter to watch him or her conduct an interview, and then watch the results of that interview on the news show. Did the reporter represent the views of the interviewee fairly? What soundbite did the TV station choose to use and why? What else should they have included on air that would have made the report more complete or fair? Report as a one-page memorandum.

5-2 Interview a television or newspaper reporter and ask about their interviewing tactics and techniques. Discuss how this knowledge will help you in the future when you are to be interviewed by a reporter. Report as a one-page memorandum.

5-3 Interview an attorney (or if your university has a law school, a third-year law student) and ask him or her to comment on the statement that "law is not about justice; it is about winning."

5-4 Attend a county court and observe both the speaking habits and body language of lawyers in front of a jury. Comment on any theatrics intended to affect the opinion of the jury and hence the outcome of the case.

5-5 Conduct mock job interviews with another student, paired by the instructor. You are an engineer interviewing the student for a position in an engineering firm. You can make up whatever you want to about the company, or use a real company. Make it as realistic as possible. You can be seated in an office and the job applicant can knock and enter, introducing himself/herself. When the interview is concluded, write a short memorandum, addressed to the head of human resources, summarizing your impression of the job candidate. Then exchange places and repeat the interview. Each of you will deliver one such memo to the professor.

CHAPTER 6

ETHICS OF ENGINEERING COMMUNICATION

- A first year engineering student has badly messed up the chemistry lab and is out of both time and reagents. The teaching assistant tells her to just copy the data from someone else and write up the lab as if she had done it herself. Should she do this?

- An engineering student, in writing a lab report, copies a technical description of the experiment from a textbook. He could have just as well written it himself but it was more convenient to copy. Should he acknowledge the source?

- An engineering student is writing a paper for his first year writing class. He asks his sophomore roommate for help and the roommate suggests a really neat topic. The first year student does not acknowledge the help of his roommate, believing that it is the writing and not the topic that is being graded.

- An engineering graduate student wishes to use an illustration from a commercially published textbook for her doctoral dissertation. Must she receive permission from the publisher?

- A client wants an engineering study on landfill location to refer to men by their last names only, not preceded by Mr., while always using Miss or Mrs. for women. The client argues that the title is a sign of respect, not gender bias. The engineer strongly disagrees, but this is what the client wants. What should the engineer do?

- An engineer writes to the newspaper asserting that another engineer who designed a failed public facility is incompetent and should be sued for damages. Is it ethical for the engineer to write such a letter? Can the project engineer sue for libel?

- A researcher finds that two out of 80 data points in a series of experiments are way off the line of best fit. She recalls that these were obtained with different reagents that may not have been pure. She decides that nobody will ever know the two data points existed, so she erases them. There is no use muddling an otherwise excellent data set. Is she doing anything unethical?

- An engineer, in writing a report for a client, borrows extensively from a previous report written by her firm for another client. Should she receive permission from the original client who has paid for the report, or should she acknowledge the "loan?"

- An engineering draftsman in a large engineering firm is working on the annual report to the stockholders and is told by a partner in the firm to truncate a scale on the growth of the firm to make it look like the performance has been outstanding. The draftsman is not asked to lie outright, but to alter the graph so the data will be presented in the most favorable light. Should he agree to redraw the graph?

In all of these cases, engineers must make decisions based on moral values. These decisions, and the processes and values that shape them illustrate the ethics of engineering communication. In this chapter we first introduce the idea of an engineering ethic and then discuss the ethics of writing and visual communication.

6.1 ETHICS IN ENGINEERING

Just as moral rules such as "tell the truth" and "don't steal" and "keep promises" apply to everyday life, similar moral rules govern the profession of engineering. These rules are spelled out in what the engineering profession calls the "Code of Ethics."

6.1.1 The code of ethics for engineers

Each engineering professional society (e.g. ASCE, ASME, AIChE, IEEE, etc.) has its own code of ethics. These codes are, however, remarkably similar. The *Code of Ethics* of the National Society of Professional Engineers (NSPE) can be used as illustrative of the engineering codes of ethics adopted by the various engineering societies. The NSPE *Code of Ethics* is in the Appendix in the back of this book.

The first section is a *Preamble* that states that engineering is a profession and therefore a code of ethics is appropriate. The *Fundamental Canons* come next:

Engineers, in the fulfillment of their professional duties, shall:
1. Hold paramount the safety, health and welfare of the public.
2. Perform services only in areas of their competence.
3. Issue public statements only in an objective and truthful manner.

148

4. Act for each employer or client as faithful agents or trustees.
5. Avoid deceptive acts.
6. Conduct themselves honorably, responsibly, ethically, and lawfully so as to enhance the honor, reputation, and usefulness of the profession.

The first canon, stating that the engineer shall hold paramount the safety, health, and welfare of the public, is the most important canon because it often overrides other obligations and requirements. The key word here is *paramount*. If there is a conflict, then this rule supersedes others.

The next section, *Section II Rules of Practice*, explains and embellishes the fundamental canons. For example, the engineer must be a whistleblower in cases where he or she feels that the public safety, health, or welfare is compromised and where no other recourse is possible. *Section III* of the Code lists a series of obligations governing engineer-engineer and engineer-client relationships.

This and other engineering codes have two types of statements that reflect morality: *admonitions* and *requirements*.

Admonitions are statements that strive to lead the engineer to the moral high ground, to make the engineer design his or her professional life so as to routinely act with moral integrity. Admonitions are statements specifying what the engineer ought to do to be a good engineer. To **not** adhere to an admonitions statement in a code of ethics will not get engineers in trouble, but to adhere to the admonition will make them better engineers. For example, in the NSPE *Code of Ethics*, guideline III-2-a reads:

a. Engineers shall seek opportunities to participate in civic affairs; career guidance for youths; and work for the advancement of the safety, health, and well-being of their community.

Nothing bad will happen (probably) if the engineer does not participate in civic affairs offering guidance for youths. The community is deprived of the engineer's technical expertise in making decisions, but the engineer cannot be reprimanded for ignoring the requirement for participating in civic affairs. This statement in the code is therefore only an admonition rather than a requirement.

Another section of the code states:

e. Engineers shall continue their professional development throughout their careers and should keep current in the specialty field by engaging in professional practice, participating in continuing education courses, reading in the technical literature, and attending professional meetings and seminars.

This sounds like another admonition, but actually it is more like a requirement, on two counts. First, not keeping up with technical developments could result in an incompetent design that may harm the public. If this happens, and the engineer is shown to be using obsolete technology, he or she is liable. The second reason is more practical. In most states renewal of the professional engineering license requires periodic updating through workshops and conferences. If an engineer wants to continue to practice professional engineering, continued professional development is a requirement.

The code of ethics also contains statement about what the engineer is required to do in order to continue to be part of the engineering community. The difference between admonitions and requirements is that not following the requirements **can** result in harm to both the engineer and the public. Ignoring the requirements would be to act immorally in professional engineering.

Here is a more explicit example of a requirement; guideline II-2-b:

> b. Engineers shall not affix their signatures to any plans or documents dealing with subject matter in which they lack competence, nor to any plan or document not prepared under their direction and control.

Suppose a civil engineer who knows little about electrical circuits approves drawings for the wiring of an elevator for a building. If the elevator fails and people get hurt, the civil engineer could not plead ignorance of electrical circuits, and this engineer would be legally (and morally) at fault.

6.1.2 Using the engineering code of ethics

When engineers are faced with having to make a decision that they perceive to have ethical dimensions, checking the code of ethics is always useful. The code is quite clear on some issues about what is and is not accepted professional practice.

For example, suppose your firm has been buying engine castings from a vendor who has an excellent record of quality and timeliness. You have no need to seek alternative sources of the castings and as far as you are concerned, the arrangement can continue forever.

One day the salesman, who you have gotten to know quite well over the years, tells you that he has tickets to the Super Bowl and asks if you and your wife would like to go. You have socialized with the salesman before and this seems like a wonderful opportunity to spend a congenial day with him. But you have an uncomfortable feeling that perhaps this is not quite right. You check the *Code of Ethics*, and find that in section 5 there is this statement:

> 5. Engineers shall not be influenced in their professional duties by conflicting interests.
>   a. Engineers shall not accept financial or other considerations, including free engineering designs, from material or equipment suppliers for specifying their product.

Based on this statement, it is quite clear that you should not accept the tickets. But why? You have no intention of changing the vendor, and he already has your business. What harm would it do to accept his Super Bowl tickets?

The reason this is not allowed is that if, for whatever reason, a problem emerges, like some of the castings are not up to code, or they cannot be delivered on time, the salesman will come to you and ask you, as a personal favor, to not penalize his company or to seek alternative vendors for the castings. You would then be in a difficult position. If you did not do your friend a favor he would view you as an ingrate. If you gave him the slack he wants, and your superiors found out about it, you would be judged (correctly) to be disloyal to your own company. In other words, you would lose either way.

So the *Code of Ethics* advises you to not accept the complimentary tickets, and for a very good reason. In this case the code is useful in keeping you out of trouble.

But you should not believe that the codes cover all ethical situations that might arise in engineering practice. The codes have, in fact, some severe limitations.

6.1.3 Limitations of the engineering code of ethics

One problem with any of the engineering codes of ethics is that the "public" to whom the engineers owe primary responsibility is not defined. This problem is especially acute for engineers working in the armaments industry. Their primary mission is to develop technology to kill the most "bad" people while protecting the "good" people. But if people are people, then are they all not part of the "public" to whom the engineer owes responsibility?

Sometimes engineers in economically wealthy countries that have strict public safety rules and regulations are tempted to ignore such regulations when doing business overseas. The manufacture and sale of products banned in the United States but legal elsewhere can cause serious ethical problems. For example, decisions such as the sale of a banned pesticide requires a definition of just which "public" the engineer is responsible to.

The engineering codes of ethics have little to say about questions regarding the environment. None of the codes spell out what, if any, responsibility engineers have to non-human animals, plants, or places. The only concern would be that the actions of the engineers not diminish the welfare of the (human) public. If an engineering project causes the demise of an animal or plant species, the concern is not for that plant or animal, but for future humans who may not be able to enjoy looking at this species or obtain some other beneficial use from it.

The conclusion we come to is that the engineering code of ethics is a fine first, but rough, tool for making ethical decisions in engineering. Often when engineers are confronted by ethical problems, a quick glance at an engineering code of ethics is enough to encourage a decision that the engineer can live with. But ethical problems are seldom straightforward, and the right actions are not obvious. There is a great deal of subtlety in ethics, and any set of guidelines such as a code of ethics cannot hope to cover all cases.

6.1.4 Making ethical decisions in engineering

In engineering, the first place to look for help is in the code of ethics. Often the first canon can squash all other considerations. If an alternative course of action does not hold **paramount** the health, safety, and welfare of the public, then quite possibly it is not the right action to take.

Second, ask some senior engineers whose opinion you value. Amazing stuff, this accumulated wisdom. They may have seen something like this before and know how it ended. They might also be able to suggest alternatives that have not occurred to you.

You need to consider all the alternatives, and try to estimate what the effect of each would have on all the people concerned. In the end, however, you have to remember that you are a professional who has moral obligations to the public and that the public depends on you to do the right thing. You are the one who has to look at yourself every morning. How would you feel if you chose a less-than-honorable option? *

6.2 MORAL CONCERNS IN ENGINEERING COMMUNICATION

Many ethicists argue that the essence of ethics is truthfulness. If we are truthful, then almost always we are bound to be acting ethically. While this may seem

---

* Being ethical is often difficult and it takes moral courage to do the right thing. In the 1930s Stalin purged (killed or imprisoned) thousands of engineers because they asked too many questions, and thus it became quite clear that engineers in Soviet Russia were expected to perform only their technical function and not ask questions about right and wrong. In that environment it is understandable that engineers simply did as they were told or ceased to practice engineering. Questioning the moral ramifications of political decisions would have resulted in an unacceptable cost.[1]

simplistic, the idea of truthfulness certainly addresses the central issue of engineering communication. Truth is usually simple and straightforward, and telling the truth is usually the safest course of action. Mark Twain advised us to "Always tell the truth. That way you don't have to remember what you said."

The negative side of truthfulness is a bit more ethically complicated. What does it mean to not tell the truth? Are all non-truths lies? Are all lies unethical?

In engineering communication, as in all communication, we find two types of non-truths: lies and deceptions. In both cases the intent is to have the recipient of the information draw a false conclusion from the available information. While both methods of eliciting false conclusions may be morally wrong, there is a well-defined operational difference between a lie and a deception.

A *lie* is a categorical statement known by the teller to be untrue. An engineer who tells a client that the client's report has been mailed, for example, while knowing full well that it is still being prepared, is telling a lie. Most lies are verbal, but lies may also be told by body language, such as the nodding of the head, or by graphics such as intentionally omitting some data points to "make the graph look better." Lying requires only that incorrect information is intentionally transmitted. If I say "I am 10 feet tall," that is a lie, regardless of whom I tell it to.* The person I tell the lie to does not have to be actively participating in the lie.

In contrast to a lie, a *deception* is an action that begins with a statement (verbal or graphical) that may be true but the intention is for the listener or reader to draw a false conclusion. The distinction between a lie and deception is that the latter requires the deceived to participate in

---

* There is an interesting extension of this principle, of course. Can one lie only to a rational adult human? Is the mere statement "I am ten feet tall" a lie if I say this to my dog? Or a rock?

reaching that false conclusion.

If the anxious client asks the engineer for the status of the report, the engineer can say that "it is essentially finished." The client might interpret this as meaning that it is being collated, in time to be picked up by FedEx that afternoon. But the fact might be that not all of it is even written. If, by using such a phrase as "essentially finished" the engineer knows that he or she is creating a false impression (without overtly lying), he or she is guilty of intentional deception, and deception with an intent to mislead or obfuscate is not honorable behavior. If the client is savvy enough to ask what the engineer means by "essentially finished," the engineer has a chance to tell the truth or to lie. The absence of the follow-up questions makes the client an active participant in the deception.

Deception in the hands of a professional such as an engineer can cause harm or have serious repercussions. For example, suppose an engineer writes a technical article about a bridge collapse and intentionally uses incomplete information that deceives the journal readership. The engineer is not lying but rather using only partial data without reporting that other data have been omitted. To publish such misleading information is unethical behavior because other engineers can draw unwarranted conclusions from the misleading report; this could result in other bridges being poorly designed, thus endangering public safety.

Lies and deceptions have moral significance because they can and often do cause harm to human beings. All people are members of a moral community, and as such, we respect each person and extend to all humans our moral concern. We have a social contract to tell the truth to each other. If all of us adhere to this principle, then all of us benefit. Imagine a world in which we can't depend on anyone telling the truth about anything. This would not be a world in which any of us would want to live. We therefore hold a shared responsibility to maintain a moral code and to conduct ourselves in an ethical manner.

Engineers have a special responsibility to act ethically, as do other professionals like physicians and lawyers, because what we do can affect the health, safety, and welfare of the public. Where such responsibility occurs, the need for truth is especially critical; ethical engineering practice requires truthfulness in all cases.

Because engineering is inherently a people-serving profession and many of the engineer's duties result in direct contact with the public, we expect engineering communications to be on a high ethical level.[3]

6.3 ETHICS AND VERBAL ENGINEERING COMMUNICATION

In engineering writing, as in other non-fiction, readers begin with the assumption that the material is:

a. true
b. written by the person identified as the author.

If engineers compromise either one of these principles, then they are either lying to or deceiving the reader.

In college, one of the most important decisions students have to make in writing is how to use outside material: when to give credit for someone else's material and when not to. And how should this be done?

Not attributing someone else's work is *plagiarism* – one of the most grievous sins in academia. Plagiarism is both stealing and deception. The person reading the material assumes that it is the author's work, and there should be no need to ask the clarifying question "Did you copy this from someone else?"

We usually think of plagiarism as verbatim reuse, without attribution, of whole works or paragraphs, but plagiarism does not have to be so blatant. Plagiarism also occurs when the writer uses key phrases without attribution. Since the key phrases form the essence of the material, such use of phrases or expressions without attribution is still plagiarism.

A third type of plagiarism occurs when the writer takes someone else's idea and puts it into different words. This is still theft of intellectual property and is morally wrong. This kind of plagiarism is the most problematic because sometimes it is difficult to distinguish your own ideas from those based on something you have read or heard. In some cases you may have read material years ago and long forgotten the source but held on to the idea. Presenting this as your own is still plagiarism even though it is unintentional.*

In whatever you write, the rule of thumb is that when in doubt, *give credit*! If you use a key phrase or helpful quotation from a book or article, place quotation marks around it and designate its source. If you do not quote directly, cite the source of the paraphrased idea.

Copying from yourself is a tricky issue. When can you use the same idea for two different papers? Can you take whole paragraphs from one paper and use them in another? As another rule of thumb, if the idea or paragraph from the old paper seems central to the theme of the new paper (if the paper would not make much sense without it) then you must acknowledge its source.

A similar problem occurs in engineering when material

---

* On a personal level, this book no doubt contains material that I, the author, have accumulated along the way during many years of teaching, and have failed to credit in this book simply because I do not know the source. I am not even sure if some of the examples and stories were mine originally or if I found them somewhere and jotted them in my class notes, intending to use them later in a lecture. In preparing this book, I tried my best to acknowledge everything. But I am sure I have missed some, in which case I am guilty of unintentional plagiarism.

within a firm is copied from old documents and sold to new clients. Should engineering firms be allowed to sell the same material over and over again to different clients without acknowledging what they are doing?

The test here is whether or not the engineer lies to or deceives the client. If the client asks the engineer whether on not the material has appeared previously in any other report, the engineer must tell the truth. If the engineer lies, then this is clearly morally wrong. In practice, most clients do not care if the material has been used before as long as the engineer has solved the problem. Most clients who frequently rely on engineers for technical advice are already familiar with how engineering is practiced and will expect that some of the material in their reports may have been used previously.

Deception in verbal engineering communication, whether by the inappropriate use of other's intellectual property or by intentionally misleading information, is antithetical to engineering practice. Sometimes it is difficult to distinguish between the right and the wrong thing to do, as illustrated by the short case studies at the beginning of this chapter. As a crude yardstick to help you make the right decision, always act as if someone is videotaping everything you do and the tape will be shown on the evening news. Would you be proud of your actions?

6.4 ETHICS AND GRAPHICAL ENGINEERING COMMUNICATION

Just as words can mislead a reader, illustrations can be deceitful. Thus, we must judge engineering illustrations not only on the value of their information and on their appearance, but also on their integrity.

In most cases, graphics have a certain sacred value to engineers and illustrations are seldom blatant lies. Few engineers and scientists will intentionally misplace a data point on a graph or add points without having the data. For this reason, most engineering readers place a great deal of weight and credibility on data points.

Sometimes, however, some researchers or practitioners fail to include all the relevant data points or will move the data points to "improve" the graph. Occasionally data points are so far off the line that the researcher is tempted to either discard them as "something went wrong here" or note them as "rogue points." There even exists a statistical test that can be used for removing (with full statistical justification) data points from a graph. This is a risky business since the one "rogue point" may in fact have been an indicator of something important and totally unexpected.

Fortunately, instances where engineers and scientists have used graphs for transmitting incorrect information (lies) seem to be rare, and in cases where there have been fabrications of data the scientific and technical community has properly condemned the actions.

Much more insidious (and more common) miscommunication of graphical information is the use of illustrations for purposes of deception. Graphs don't have to actually lie to express incorrect information, since it is the *perceived* information that matters.

Most misleading illustrations are deceptions rather than lies. The person receiving the information is an active participant in the transfer of information and draws intentionally unwarranted conclusions. Such deception can be achieved by several unethical techniques, including inappropriate cause and effect, unwarranted visual embellishment, and misuse of data.

6.4.1 Implied cause and effect

The first type of graphical deception is implied cause and effect. For example, Figure 6-1 shows the correlation between the rate of typhoid fever deaths and the fraction of

160

the population with public water supplies. Such graphs have been repeatedly published by environmental engineers who want to convince the world that they have performed magnificently and should receive due credit for their efforts. There is clearly no doubt that the construction of clean public water supplies helped reduce the typhoid death rate, *but this graph does not prove the cause and effect relationship*. An excellent correlation also results when the typhoid death rate is plotted against some totally unrelated set of data such as the increase in the manufacture of automobile tires. The conclusion, if causation is mistaken for correlation, is that either automobile tires reduce typhoid deaths or the reduction in typhoid deaths resulted in the increase in tires.

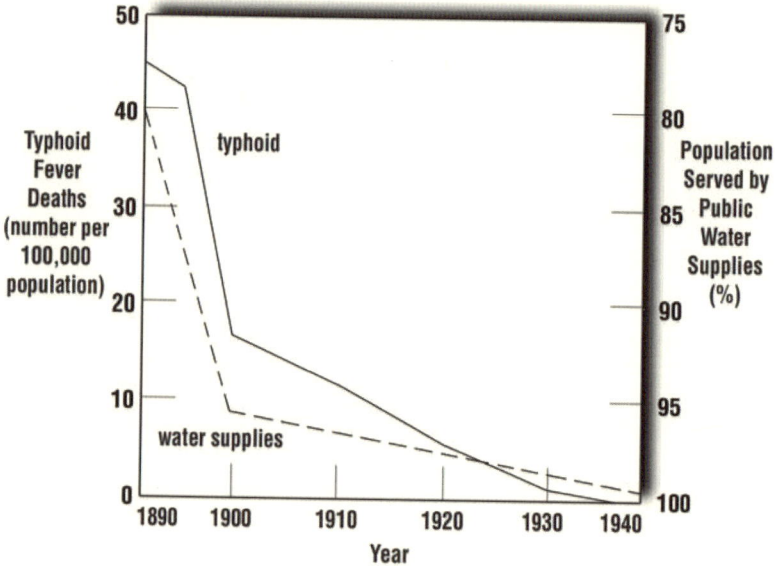

Figure 6-1 Correlation between population with public water supplies and the typhoid fever death rate in the United States.[2]

Some graphs may have been constructed with the best of intentions, but the resulting graphic may be deceptive. For example, a consulting engineering firm was asked to study the feasibility of recycling for an unnamed community. The author of the study found that some cities had successful programs and others were not quite so successful, and that there was a curious correlation between the total time the program had been in operation and the success of the program as measured by the rate of participation. The resulting graphic is shown below.

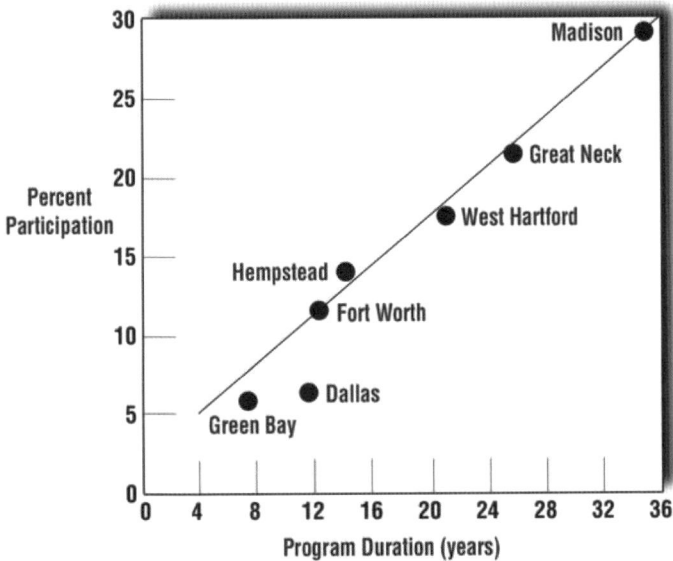

Figure 6-2 Correlation between the percent participation and the duration of recycling programs in various cities.

The implication is that if one wants to have a successful program, all one has to do is to have it run for a long time. But this is terribly misleading. The program in Madison had existed for three years **because** it was

162

successful. The folks in Madison were enthusiastic recyclers and thus the program was successful. The graphic can easily be misinterpreted and is, perhaps unintentionally, deceptive.

6.4.2 Visual Embellishments

A second common source of misperceived and thus unethical graphics is the use of visual embellishments. One common visual embellishment technique is the two-dimensional bar graph. A bar in a bar graph is one-dimensional, showing the quantity as its height. But the bar can also be shown as a two-dimensional picture. Figure 6-3 shows how two-dimensional trash cans can represent one-dimensional bars. From this figure we see that even though 4.5 is only 50 percent larger than 3.0, the appearance is much greater because the reader sees the *area* of the trash cans, not just the height.

Solid Waste Generation
1960   1990
3.0 lb/cap/day   4.5 lb/cap/day

Figure 6-3 Deceptive use of two-dimensional pictures to represent one-dimensional data.

6.4.3 Misrepresentation of Data

Some graphs are clearly intended to mislead. For example, Figure 6-4 shows a notorious graph that appeared in a government document arguing that the United States was losing its edge in science and technology, since our share of the Nobel prizes seemed to have dropped precipitously. The truth is that the last data point (1971-1974) is the total number of Nobel prizes for only a *five year interval*, whereas all the remaining points represent Nobel prizes received during *ten year intervals*.

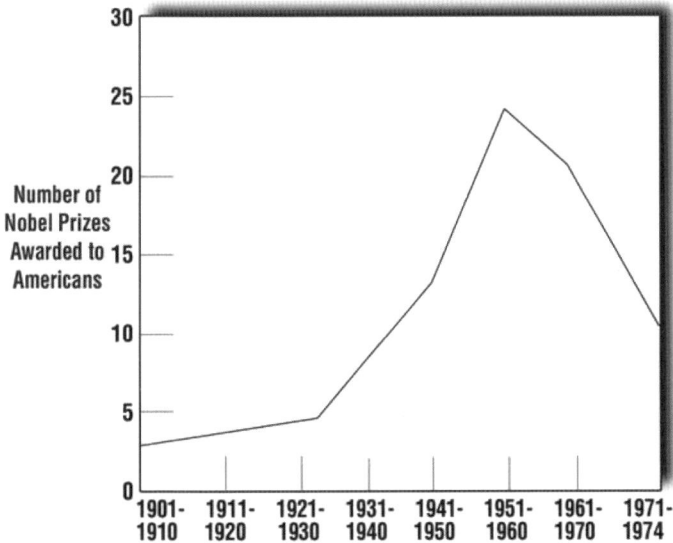

Figure 6-4 Misleading line graph due to uneven scale. [3]

Some graphs deceive because the data are plotted cumulatively and the reader is not sufficiently warned to interpret the graph in such fashion. Figure 6-5 shows the use of various forms of power for electricity production. A quick glance suggests that nuclear power is the most

164

important energy source and continues to provide the greatest share of electricity. The data in this graph are, however, plotted cumulatively so that the energy production from various sources is the difference between the adjacent lines.

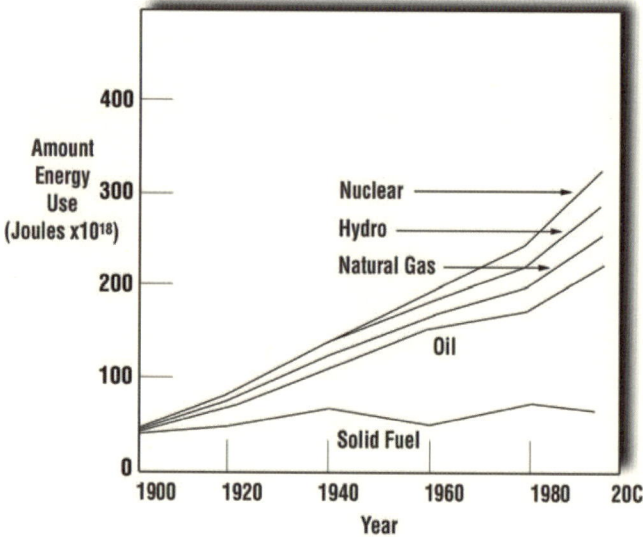

Figure 6-5 An example of a deceptive graph because the data are summed.

Broken scales can also convey misleading information. Broken scales occur either when an axis does not start at zero or when the scale is temporarily discontinued. Note how the top graph in Figure 6-6 creates a mistaken impression by not starting at zero and thus conveys a very different impression from the more honest graph in Figure 6-7. Good practice requires that all broken scales be clearly indicated. For example, Figures 3-1 and 3-2 in Chapter 3 show broken scales that are clearly drawn.

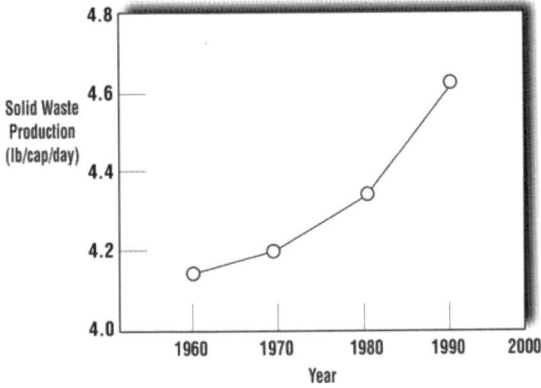

Figure 6-6 Deceptive graph due to inappropriately broken scale.

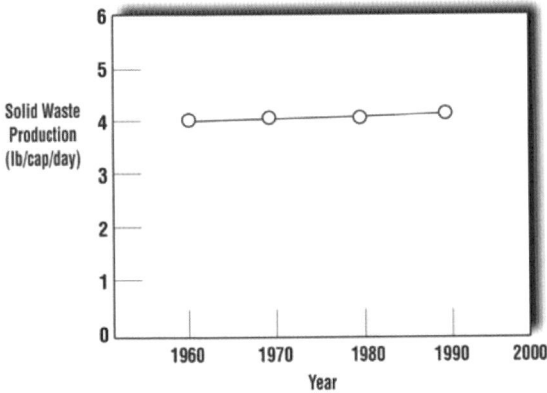

Figure 6-7 Same graph as Figure 6-6 without the broken scale, leaving a very different impression.

Sometimes it is possible to conceal data points in such a way as to make the rest of the data appear much more

166

impressive and thereby mislead the reader. Like Figure 6-8, Figure 6-9 shows accurate points, but the abscissa is moved so that the points nestle up against the ordinate axis thus concealing them and creating a false impression of a correlation that may not exist.

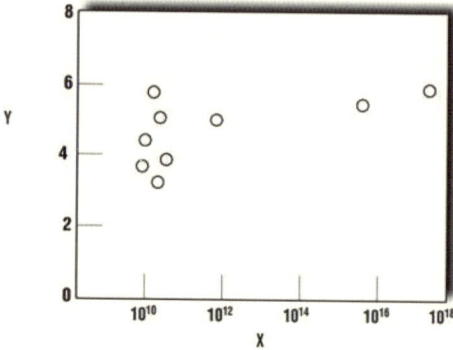

Figure 6-8 Data as obtained in a research experiment.

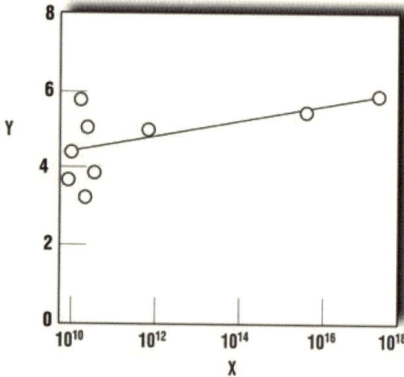

Figure 6-9 Same data as in Figure 6-7, replotted to misrepresent the results.

Sometimes the authors of a graph read more into the data than a reasonable person might. Figure 6-10 shows actual data used as the foundation for a complex theoretical model. What should be the line? Try to imagine where you would draw the line of best fit.

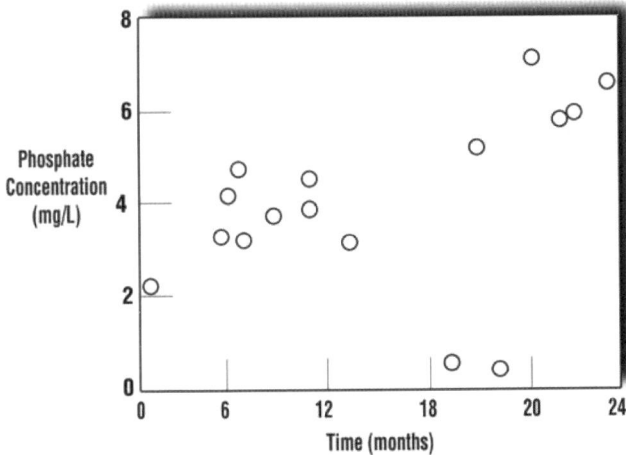

Figure 6-10 Data obtained in an actual research experiment.

Figure 6-11 shows the same data with the line as drawn by the authors. There is no reason to draw such a line, except that the theoretical model developed by the authors predicted that this line would occur. The most that can be said in such cases is that the data certainly did not *dis*prove the model. Claiming that the data offer proof of the model is, however, clearly unwarranted.

168

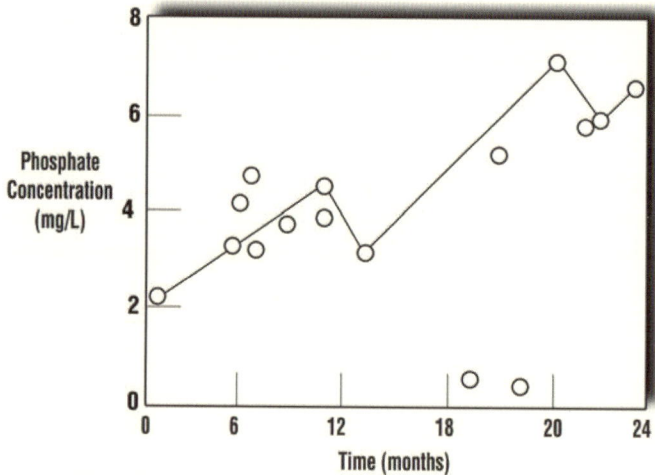

Figure 6-11 The same data as in Figure 6-10, with the line drawn in by the authors.

6.4.4 Misuse of statistics

Statistics* can be used to suggest causation where none exists. The famous British prime minister Benjamin Disraeli is supposed to have said that there are three kinds of lies: lies, damn lies, and statistics. Disraeli notwithstanding, as long as statistics are calculated accurately the result cannot represent lies. On the other hand, statistics opens up a tremendous opportunity to deceive.

---

* The word *statistics* can be singular or plural. If statistics refers to the science of calculating probability (e.g. like other sciences such as biology or psychology), statistics is written with an "s" at the end of the word but treated as a singular word. A *statistic* is one of many tools statisticians use to understand probability, such as mean, standard deviation, mode, etc. If there are several of these, then they are *statistics* (plural).

One of the most used (and abused) statistical techniques is the least squares fit of data. The theory is that the best fit is obtained when the sum of the squares of the vertical distances between the experimental values of Y and the line representing the relationship between X and Y is minimized. If the sum of the square of these distances approaches zero, then the fit is perfect.

The statistic used to measure this goodness of fit is known as R^2. If the calculated $R^2 = 1.0$, then the data fit perfectly. That is, we get a perfect straight line and all of the data points fall exactly on the line. If $R^2 = 0$, then there is no fit whatsoever. That is, there is no correlation between X and Y. Then the values of Y are random with regard to X, and X cannot be used to predict the values of Y.

Shady ethics enter when R^2 is used to determine the goodness of fit without also including the plotted data, or without using common sense. A powerful example of how statistical correlation can be subverted was published by Anscombe.[4] Figure 6-12 shows how four data sets can be plotted to yield exactly the same $R^2 = 0.67$. Each data set has the same mean and is described by the same least squares equation. And yet the data, when plotted and viewed on the plots, represent four totally different populations. Only by plotting the data would the reader understand the actual relationship between the two variables.

6.5 TRUTHFULNESS IN ENGINEERING COMMUNICATION

Every engineering office or department has an "engineer in responsible charge" of projects. Their job is to make sure that their project is completed successfully and within budget. This engineer in responsible charge places his or her professional engineering seal on design drawings or final reports in order to certify that everything is accurate

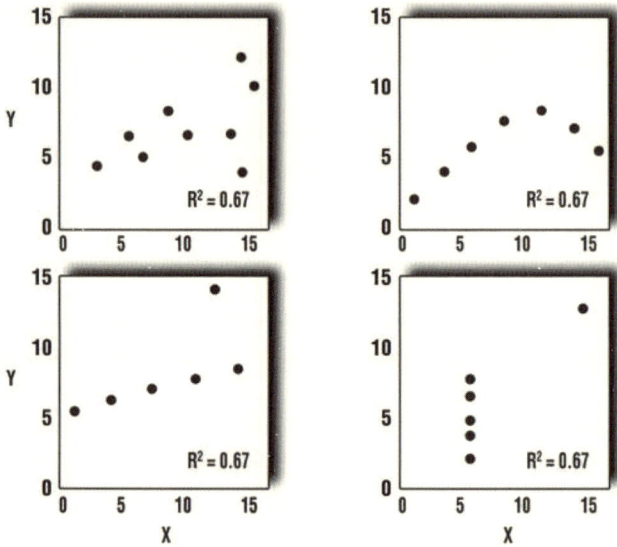

Figure 6-12 Four plots with data showing identical R^2 values. [4]

and complete, and that whatever has been designed will work as specified.[*] In sealing drawings or otherwise accepting responsibility, the engineer in charge places his or her professional integrity and professional honor on the line. The engineer in charge can not hide if something goes wrong. If something *does* go wrong, "One of my younger engineers screwed up," is not a defense, because the engineer in charge is supposed to have overseen the calculations or the design.

For very large projects where the responsible engineer may not even know all of the engineers working on the project, much less be able to oversee their calculations, this is clearly impossible. In a typical engineering office the

---

[*] It has been rumored that in some European countries, not too many years ago, the engineer in responsible charge of building a bridge was actually required to stand under the bridge while it was tested for bearing capacity!

responsible engineer depends on a team of senior engineers who oversee other engineers, who oversee others and so on down the line. How can the responsible engineer at the top of the pyramid be confident that the product of collective engineering skills meets the client's requirements?

Fortunately, the rules governing this activity are fairly simple. Central to these rules is the concept of truthfulness in engineering communication. *Technical communication up and down the organization requires an uncompromising commitment to tell the truth no matter what the next level of engineering wants to hear.*

It is theoretically possible for an engineer in the lower ranks to develop spurious data, lie about test results, or generally manipulate the basic design components. Such information might not be readily detected by supervisory engineers if the bogus information is beneficial to the completion of the project. If the information is not beneficial, on the other hand, everyone along the chain of engineering responsibility will give it a hard critical look. Therefore the inaccurate information, if it is the desired information, can steadily move up the engineering ladder because *at every level the tendency is not to question good news.* The superiors at the next level also want good news, and want to know that everything is going well with the project. They do not want to know that things may have gone wrong somewhere at the basic level. Knowing this, and fearing being shot as the messenger, engineers tend to accept good news and question bad news. In short, the axiom that "good news travels up the organization – bad new travels down," holds for engineering as well. And bad news will travel down very quickly (perhaps in the form of: "You're fired!")

The only correcting mechanism in engineering exists at the very end of the project if failure occurs: the software crashes, the bridge falls down, the project is grossly overbid, or the refinery explodes. And then the search begins for what went wrong. Eventually the truth emerges, and often the problems can be traced to the initial level of engineering design, the development of data, and the interpretation of test results.

For this reason engineers, especially young engineers, must be extremely careful in their work. It is one thing to make a mistake (we all do), but it is another thing totally to use misinformation in the design. Fabricated or spurious test results can lead to catastrophic failures when the only detection mechanism is at the conclusion of the project.

It will be some time until you the engineering student will find yourself in "responsible charge," and you may own a professional engineering seal for years before you first use it to seal drawings. But when you do, that first time you place your name, reputation, and career on the line, you want to know that all of the engineers who worked on this project did their very best engineering and at the absolute minimum, were truthful. Without such trust in engineering, the system will fail. It is your responsibility now as an engineering student to do your part in the engineering enterprise by being absolutely truthful in all you do. If I may paraphrase Bronowski ...[5]

All engineering projects are communal; there would be no computers, there would be no airplanes, there would not even be civilization, if engineering was a solitary activity. What follows? It follows that we must be able to rely on other engineers; we must be able to trust their work. That is, it follows that there is a principle which binds engineering together, because without it the individual engineer would be helpless. This principle is truthfulness.

173

REFERENCES

173

REFERENCES

1. Graham, L., *The Ghost of the Executed Engineer*, Harvard University Press, Cambridge MA, 1996.
2. After Whipple and Horwood, from Fair, G. M., J. C. Geyer and D. A. Okun, *Water and Wastewater Engineering,* John Wiley & Sons, New York 1966.
3. National Science Foundation, *Science Indicators, 1974* Washington, DC, 1976.
4. Anscombe, F. J., "Graphs in Statistical Analysis," *Am. Statistician*, 27, 1973, 17-21; as described in Betthouex, P., M. and L. C. Brown, *Statistics for Environmental Engineers,* Lewis Publishers, Boca Raton, FL, 1994.
5. Quoted by Ian Jackson, *Honor in Science,* Sigma Xi, 1894, p.7, from J. Bronowski, *Science and Human Values,* Messner, New York, 1956 p.73.

PROBLEMS

6-1 Choose a passage of about 100 words from one of your favorite books. Plagiarize this passage in two ways:

a) Rewrite the passage by lifting whole phrases so that it appears to be in your own writing but in fact is plagiarized.
b) Use the central idea from the passage and express it in your own words but without using any of the key phrases verbatim. Include a photocopy of the original passage with the plagiarized paragraphs.

6-2 Choose any graphic from a national magazine or newspaper. Analyze the graphic in terms of its truthfulness. Does it express the intended idea with accuracy? If not, why not? Redraw the graphic to reflect a *biased* view. In other words, change the graphic so that it is deceitful. Alternatively, of you find a biased graphic, revise it to

174

make it truthful. Submit both the original and the doctored
versions. Explain the differences.

6-3. Prepare a (minimum) two page discussion on any one
of the ethical dilemmas used to introduce this chapter. Use
the rubric presented in paragraph 6.1.4 to analyze the
dilemma. Be sure to reflect on the values involved. What
are the possible courses of action, and what might be the
repercussions of these actions? How would *you* handle the
situation if you had to make the decision? Be prepared to
discuss your response in class and to defend your views.

6-4. The paragraphs below were taken from Imsonovich,
E. I. (1996) "Ethics and Rationality," *Journal of Weird
Ethical Argument*, 9(13), 1-12.

Most of us would choose not to rob banks, and
would look with disfavor upon people who do. We are
tempted to say that such an act by another person is an
irrational act, especially if the chances of getting caught
are high. But people choose to rob banks every day, and
to them at the moment they are committing the robbery,
this is not an irrational act. If we are tempted to call an
act by someone else irrational, it simply means that we
cannot understand fully all of the beliefs carried in that
person's head that made him or her decide to rob the
bank. To that person, at the time he or she is holding up
the bank, it is not an irrational act.

I may choose to rob a bank myself, and get caught.
Sitting in my cell, during what some psychologists call
the "cool moment," I may look back at the decision to
rob the bank as an irrational act. I wish I had not done
it. I may even begin to think of it as an immoral thing to
do (or I may even get religion), and may sincerely
regret my actions. I will see my past actions as
irrational. But I have to admit that this evaluation is

only possible from a different time. At the time the act occurred, it was not an irrational act.

Thus an act can be irrational only if it is removed in space or time from the actual act. We can think of someone else's act as irrational (we don't believe we would have done it) or we can think of our own actions as irrational either before or after the act has occurred. In the latter case, irrational simply means that viewing the act from our present time, space and circumstance, we would not have acted in such a way. "What was I thinking of?" is the usual question when we contemplate what we have done. Unfortunately, we can never recreate the exact combination of beliefs at that time that led us to the now regretful action. Nevertheless, at the time the action occurred, it was a rational act, *or we would not have done it*!

The conclusion one must therefore draw is that there are no irrational acts at the time the acts are performed. And if there are no *irrational* acts, there cannot be *rational* acts. The world "rational" loses its meaning, and completely negates 2500 years of philosophical thought since all normative philosophy is based on the action of the "rational person."

Plagiarize this passage by summarizing it in your own words, but without using any of the original key expressions.

6-5. Study graphics used in newspapers or magazines and collect examples of common methods of deceitfulness (optical illusion, implied causation, visual embellishments, and misuse of data). Select a few of these and write down why you believe the graphics are deceitful. Submit the original copies and your analyses.

6-6 The following test scores were obtained on an examination:

Student	height	gender	score
1	5 ft. 3 in.	male	49
2	6 ft. 0 in.	male	87
3	4 ft. 9 in.	female	99
4	5 ft. 2 in.	female	49
5	5 ft. 0 in.	female	79
6	5 ft. 10 in.	male	81
7	5 ft. 11 in.	male	88
8	5 ft. 6 in.	female	100
9	4 ft. 10 in.	female	77
10	5 ft. 11 in.	male	82
11	4 ft. 11 in.	female	60
12	5 ft. 6 in.	female	73

Draw *biased* graphics showing the superiority of
    a) tall persons
    b) females

6-7 Use the data shown in Problem 3-9 to draw a *biased* graphic. Explain why and how this is deceitful.

6-8 Consider the scenario below. If an engineer writes a paper for a trade journal or a technical journal, is the engineer writing as an individual or as a member of a firm? Does the paper then belong to the engineering firm, or to the individual? What should Jeff Henry do?

    Jeff Henry is highly regarded in the structural engineering world and is often sought out to perform high profile jobs such as bridge failures. His firm, Riveters Engineering, is prosperous and employs about 20 engineers, among them Miguel Hernandez.

Henry did a particularly interesting piece of forensic engineering on the failure of the Twiggy Narrows Bridge and published the results in *Civil Engineering*, the flagship journal of the civil engineering profession.

A few months later, Henry is leafing through a recent copy of *The Consulting Engineer* and finds an article by Miguel Hernandez. His immediate reaction is pride in his younger colleague in having the initiative to publish such an article. It will no doubt enhance the already excellent reputation of the firm. The article talks about the science of forensic engineering in structural failures and uses as examples several of the projects Henry has worked on. In particular, the Twiggy Narrows Bridge collapse is discussed, and as Henry is reading, the prose is sounding terribly familiar. He takes out his own article and compares the text. It is identical. Hernandez has simply taken Henry's article and reprinted it as part of his own.

Henry confronts Hernandez with this and Hernandez is dumbfounded.

"Of course I copied it. Aren't we in the same firm? Don't we routinely copy each other's drawings and calculations? If one of us designs a certain connector for a truss, why go through another calculation if the same connector is used for another job? Who did the original calculation is unimportant. Why are you so upset that I used a part of your article for the other article? It's all for the good of the firm, isn't it?"

6-9 Consider the scenario below. What rights did Gene McJunkin have to the data he developed for his client? Did he lose all ownership of the data the moment he wrote the report for his client? Does he now have an obligation to do something, or should he be, as the NSPE *Code of Ethics* advises, a "faithful agent of his client" and keep quiet?

The Jordan Textile Company has been advised by the State Environmental Management Authority that it has 60 days to respond to a letter charging the company with causing water quality problems as a result of their industrial waste being discharged into the Haw River.

Jordan Textiles hires Gene McJunkin to study the matter. After extensive testing of the stream and sampling of the industrial wastewater, McJunkin concludes that if the effluent sampling is done at appropriate times of the day, when the vats are not dumped, then the discharge would meet the NPDES requirements. Since the discharge from the manufacturing plant is intermittent, the wastewater treatment facility is sometimes overloaded and cannot meet the present permit requirements, most likely causing the periodic fish kills and other water quality problems. None of these problems, however, is directly related to human health, and only by doing a careful sampling study would these intermittent discharges be discovered. McJunkin also concludes that if the industrial discharge is to be treated adequately, the old plant would have to be scrapped and a new facility built, at a cost of several million dollars.

McJunkin delivers a preliminary written report to his client, Jordan Textiles. The company management appears to be satisfied with the preliminary report and he is asked not to prepare a final report. He is fully compensated for his services and his contract is terminated.

The State decides to hold a public hearing on the tightening of the discharge permit for Jordan Textiles. Gene McJunkin, out of curiosity, goes to the meeting and is dumbfounded to hear the engineer at Jordan Textiles present selective data to try to convince the State that the discharge from Jordan Textiles cannot be responsible for the fish kills. McJunkin knows that this is not so. His data proved that Jordan Textile was most probably responsible for the fishkills. But the Jordan engineer is using selected data to "prove" otherwise. McJunkin catches the eye of the Jordan engineer, who immediately looks away.

6-10 Consider the scenario below. What rights does Saam El-Hatem have in this conflict? He wrote the report. Does it then belong to him or the firm, or to his clients? What should he do?

Large consulting firms commonly have many offices, and often communication among the offices is less than efficient. Saam El-Hatem, who works in the Atlanta office, is retained by a neighborhood association to write an environmental impact study that concludes that the plans by a private oil company to build a petrochemical complex would harm the habitat of an endangered species. The client, the neighborhood association, has already reviewed draft copies of the report and is planning to hold a press conference when the final report is delivered. El-Hatem, as the author of the report, is asked to attend the news conference.

Engineer Bruce Cabot, a partner in the firm and working out of the New York office, receives a phone call.

"Cabot, this is J. C. Octane, president of Bigness Oil Company. As you well know, we have retained your firm for all of our business and have been quite satisfied with your work. There is, however, a minor problem. We are intending to build a refinery in the Atlanta area and hope to use your company for the design."

"We would be pleased to work with you again," replies Cabot, already counting the $1 million design fee.

"There is, however, a small problem," continues J. C. Octane. "It seems that one of your engineers in the Atlanta office has conducted a study for a neighborhood group opposing our refinery. I have received a draft copy of the study, and my understanding is that the engineer and leaders of the neighborhood organization are to hold a press conference in a few days and conclude unfavorable environmental impact as a result of the refinery. I need not tell you how disappointed we will be if this occurs."

As soon as Cabot hangs up the phone with J. C. Octane he calls El-Hatem in Atlanta.

"Saam, you must postpone the press conference at all costs," Cabot yells into the phone.

"Why? It's all ready to go," counters El-Hatem.

"Here's why. You had no way of knowing this, but Bigness Oil is one of the firm's most valued clients. The president of the oil company has found out about your report and threatens to pull all of their business should the report be delivered to the neighborhood association. You have to rewrite the report in such a way as to show that there would be no significant damage to the environment."

"I can't do that!" pleads El-Hatem.

"Let me see if I can make it clear to you then," replies Cabot. "You either rewrite the report or withdraw from the project and write a letter to the neighborhood association stating that the draft report was in error, and offer to refund all of their money. You have no other choice!"

Appendix

National Society of Professional Engineers

Code of Ethics for Engineers

Preamble

Engineering is an important and learned profession. As members of this profession, engineers are expected to exhibit the highest standards of honesty and integrity. Engineering has a direct and vital impact on the quality of life for all people. Accordingly, the services provided by engineers require honesty, impartiality, fairness, and equity, and must be dedicated to the protection of the public health, safety, and welfare. Engineers must perform under a standard of professional behavior that requires adherence to the highest principles of ethical conduct.

I. Fundamental Canons

Engineers, in the fulfillment of their professional duties, shall:

1. Hold paramount the safety, health and welfare of the public.
2. Perform services only in areas of their competence.
3. Issue public statements only in an objective and truthful manner.
4. Act for each employer or client as faithful agents or trustees.
5. Avoid deceptive acts.
6. Conduct themselves honorably, responsibly, ethically, and lawfully so as to enhance the honor, reputation, and usefulness of the profession.

II. Rules of Practice

1. Engineers shall hold paramount the safety, health, and welfare of the public.

a. If engineers' judgment is overruled under circumstances that endanger life or property, they shall notify their employer or client and such other authority as may be appropriate.

b. Engineers shall approve only those engineering documents that are in conformity with applicable standards.

c. Engineers shall not reveal facts, data, or information without the prior consent of the client or employer except as authorized or required by law or this Code.

d. Engineers shall not permit the use of their name or associate in business ventures with any person or firm that they believe are engaged in fraudulent or dishonest enterprise.

e. Engineers shall not aid or abet the unlawful practice of engineering by a person or firm.

f. Engineers having knowledge of any alleged violation of this Code shall report thereon to appropriate professional bodies and, when relevant, also to public authorities, and cooperate with the proper authorities in furnishing such information or assistance as may be required.

2. Engineers shall perform services only in the areas of their competence.

a. Engineers shall undertake assignments only when qualified by education or experience in the specific technical fields involved.

b. Engineers shall not affix their signatures to any plans or documents dealing with subject matter in which they lack competence, nor to any plan or document not prepared under their direction and control.

c. Engineers may accept assignments and assume responsibility for coordination of an entire project and sign and seal the engineering documents for the entire project, provided that each technical segment is signed and sealed only by the qualified engineers who prepared the segment.

3. Engineers shall issue public statements only in an objective and truthful manner.

a. Engineers shall be objective and truthful in professional reports, statements, or testimony. They shall include all relevant and pertinent information in such reports, statements, or testimony, which should bear the date indicating when it was current.

b. Engineers may express publicly technical opinions that are founded upon knowledge of the facts and competence in the subject matter.

c. Engineers shall issue no statements, criticisms, or arguments on technical matters that are inspired or paid for by interested parties, unless they have prefaced their comments by explicitly identifying the interested parties on whose behalf they are speaking, and by revealing the existence of any interest the engineers may have in the matters.

4. Engineers shall act for each employer or client as faithful agents or trustees.

a. Engineers shall disclose all known or potential conflicts of interest that could influence or appear to influence their judgment or the quality of their services.

b. Engineers shall not accept compensation, financial or otherwise, from more than one party for services on the same project, or for services pertaining to the same project, unless the circumstances are fully disclosed and agreed to by all interested parties.

c. Engineers shall not solicit or accept financial or other valuable consideration, directly or indirectly, from outside agents in connection with the work for which they are responsible.

d. Engineers in public service as members, advisors, or employees of a governmental or quasi-governmental body or department shall not participate in decisions with respect to services solicited or provided by them or their organizations in private or public engineering practice.

e. Engineers shall not solicit or accept a contract from a governmental body on which a principal or officer of their organization serves as a member.

5. Engineers shall avoid deceptive acts.

a. Engineers shall not falsify their qualifications or permit misrepresentation of their or their associates' qualifications. They shall not misrepresent or exaggerate their responsibility in or for the subject matter of prior assignments. Brochures or other presentations incident to the solicitation of employment shall not misrepresent pertinent facts concerning employers, employees, associates, joint venturers, or past accomplishments.

b. Engineers shall not offer, give, solicit or receive, either directly or indirectly, any contribution to influence the award of a contract by public authority, or which may be reasonably construed by the public as having the effect of intent to influencing the awarding of a contract. They shall not offer any gift or other valuable consideration in order to secure work. They shall not pay a commission, percentage, or brokerage fee in order to secure work, except to a bona fide employee or bona fide established commercial or marketing agencies retained by them.

III. Professional Obligations

1. Engineers shall be guided in all their relations by the highest standards of honesty and integrity.

a. Engineers shall acknowledge their errors and shall not distort or alter the facts.

b. Engineers shall advise their clients or employers when they believe a project will not be successful.

c. Engineers shall not accept outside employment to the detriment of their regular work or interest. Before accepting any outside engineering employment they will notify their employers.

d. Engineers shall not attempt to attract an engineer from another employer by false or misleading pretenses.

e. Engineers shall not promote their own interest at the expense of the dignity and integrity of the profession.

2. Engineers shall at all times strive to serve the public interest.

a. Engineers shall seek opportunities to participate in civic affairs; career guidance for youths; and work for the advancement of the safety, health, and well-being of their community.

b. Engineers shall not complete, sign, or seal plans and/or specifications that are not in conformity with applicable engineering standards. If the client or employer insists on such unprofessional conduct, they shall notify the proper authorities and withdraw from further service on the project.

c. Engineers shall endeavor to extend public knowledge and appreciation of engineering and its achievements.

3. Engineers shall avoid all conduct or practice that deceives the public.

a. Engineers shall avoid the use of statements containing a material misrepresentation of fact or omitting a material fact.

b. Consistent with the foregoing, engineers may advertise for recruitment of personnel.

c. Consistent with the foregoing, engineers may prepare articles for the lay or technical press, but such articles shall not imply credit to the author for work performed by others.

4. Engineers shall not disclose, without consent, confidential information concerning the business affairs or technical processes of any present or former client or employer, or public body on which they serve.

a. Engineers shall not, without the consent of all interested parties, promote or arrange for new employment or practice in connection with a specific project for which the engineer has gained particular and specialized knowledge.

b. Engineers shall not, without the consent of all interested parties, participate in or represent an adversary interest in connection with a specific project or proceeding in which the engineer has gained particular specialized knowledge on behalf of a former client or employer.

5. Engineers shall not be influenced in their professional duties by conflicting interests.

a. Engineers shall not accept financial or other considerations, including free engineering designs, from material or equipment suppliers for specifying their product.

b. Engineers shall not accept commissions or allowances, directly or indirectly, from contractors or other parties dealing with clients or employers of the engineer in connection with work for which the engineer is responsible.

6. Engineers shall not attempt to obtain employment or advancement or professional engagements by untruthfully criticizing other engineers, or by other improper or questionable methods.

a. Engineers shall not request, propose, or accept a commission on a contingent basis under circumstances in which their judgment may be compromised.

b. Engineers in salaried positions shall accept part-time engineering work only to the extent consistent with policies of the employer and in accordance with ethical considerations.

c. Engineers shall not, without consent, use equipment, supplies, laboratory, or office facilities of an employer to carry on outside private practice.

7. Engineers shall not attempt to injure, maliciously or falsely, directly or indirectly, the professional reputation, prospects, practice, or employment of other engineers. Engineers who believe others are guilty of unethical or illegal practice shall present such information to the proper authority for action.

a. Engineers in private practice shall not review the work of another engineer for the same client, except with the knowledge of such engineer, or unless the connection of such engineer with the work has been terminated.

b. Engineers in governmental, industrial, or educational employ are entitled to review and evaluate the work of other engineers when so required by their employment duties.

c. Engineers in sales or industrial employ are entitled to make engineering comparisons of represented products with products of other suppliers.

8. Engineers shall accept personal responsibility for their professional activities, provided, however, that engineers may seek indemnification for services arising out of their practice for other than gross negligence, where the engineer's interests cannot otherwise be protected.

a. Engineers shall conform with state registration laws in the practice of engineering.

b. Engineers shall not use association with a nonengineer, a corporation, or partnership as a "cloak" for unethical acts.

9. Engineers shall give credit for engineering work to those to whom credit is due, and will recognize the proprietary interests of others.

a. Engineers shall, whenever possible, name the person or persons who may be individually responsible for designs, inventions, writings, or other accomplishments.

b. Engineers using designs supplied by a client recognize that the designs remain the property of the client and may not be duplicated by the engineer for others without express permission.

c. Engineers, before undertaking work for others in connection with which the engineer may make improvements, plans, designs, inventions, or other records that may justify copyrights or patents, should enter into a positive agreement regarding ownership.

d. Engineers' designs, data, records, and notes referring exclusively to an employer's work are the employer's property. The employer should indemnify the engineer for use of the information for any purpose other than the original purpose.

e. Engineers shall continue their professional development throughout their careers and should keep current in their specialty fields by engaging in professional practice, participating in continuing education courses, reading in the technical literature, and attending professional meetings and seminars.

INDEX